No Metal No Magic

Element 17

Chlorine, Presented By Krystix, From The Magical Elements of the Periodic Table Book Series

17   35.45

Cl

chlorine

Krystix

Chlorine

By Sybrina Durant with Illustrations by Pranavva et al.

Chlorine, Presented By Krystix, From The Magical Elements of the Periodic Table Book Series

Story copyright 2026

Soft Cover Print ISBN 13 — 978-1-942740-64-3

BISAC Codes:

JNF051070 JUVENILE NONFICTION / Science & Nature / Chemistry

JNF016000 JUVENILE NONFICTION / Curiosities & Wonders

JNF051080 JUVENILE NONFICTION / Science & Nature / Earth Sciences / General

Krystix Presents Chlorine

This Element 17 book features the periodic table element, Chlorine. It is presented by Krystix, member of the Elemental Dragon Clan. Each dragon has a magical tail tipped with an element that gives them unique powers. Their powers are based on the properties of their element.

Antz is just one of the 118 elementals who will present all of the Magical Elements of the Periodic Table to readers who are curious about the wonders of the world.

Krystix introduces the very magical element, Chlorine, in his book.

The Elemental Dragon Clan and their other techno-magical friends are the perfect group to introduce you to metals and other elements in the Periodic Table. Hopefully, this Magical Element of the periodic table book will spark an interest in the magical and real world properties of all the metals and other elements known today. You may be surprised at how prominently they feature in our every day lives.

Each page in this book contains terms that might not be completely familiar to the reader. Refer to the definitions in the back of the book to get a clear understanding of each meaning.

There is also a fun elemental themed Periodic Table at the back of the book. It features 118 elements presented by fanciful characters like unicorns, dragons, wizards, knights and goblins.. They want you to remember that if there's no metal...there's no magic or technology.

Remember, "No metal – No Magic. . .and No Technology".

It's Techo-Magical.

Note: Sybrina Publishing websites are Sybrina.com and MagicalPTElements.com. Follow sybrinapublishing on Instagram, Magical Elements of the Periodic Table on Facebook, @sybrinad on Pinterest, Sybrina_SPT on Twitter; and Sybrina Durant on LinkedIn.

Chlorine is a Halogen

Chlorine (Cl) was first recognized as a gaseous substance in 1630 in Brussels, Belgium. The true nature of Chlorine gas as an element was recognized in 1810 by English chemist Humphry Davy, who later named it Chlorine. The Greek word for "greenish-yellow" is khloros.

Chlorine is the 2nd lightest Halogen gas. It is greenish-yellow and has a very pungent smell.

Chlorine is a non-metal element that acts as an electrical insulator in its pure gas, liquid, or solid forms. Because its electrons are tightly bound in covalent bonds rather than free to move, it cannot conduct electrical current. It is Diamagnetic and does not conduct electricity in pure forms.

Chlorine is a highly reactive gas that readily bonds with metals and non-metals to form stable compounds.

Chlorine is a Halogen, specifically classified in Group 17 of the periodic table. As a highly reactive nonmetal, it sits between fluorine and bromine.

LEGEND

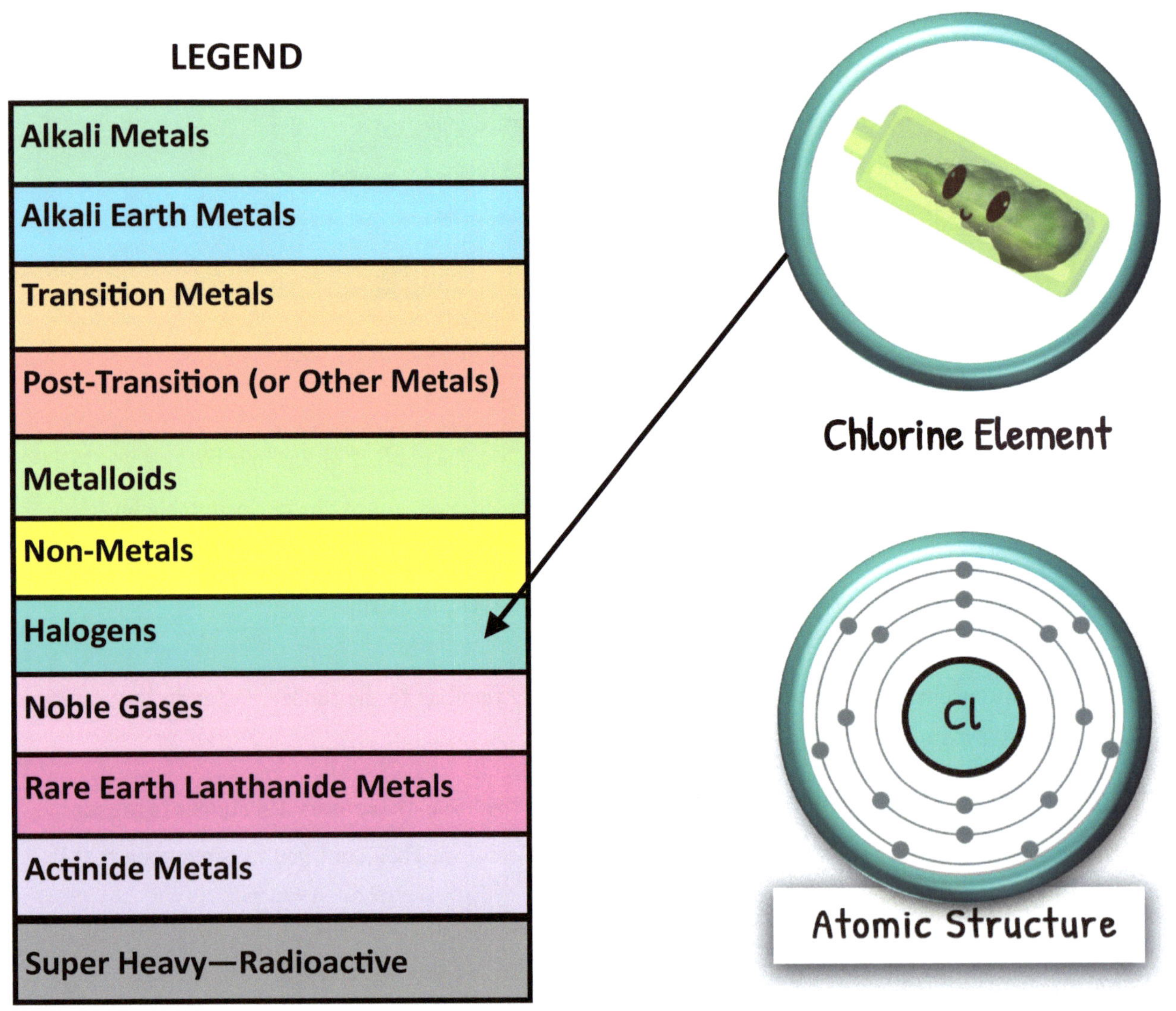

Alkali Metals

Alkali Earth Metals

Transition Metals

Post-Transition (or Other Metals)

Metalloids

Non-Metals

Halogens

Noble Gases

Rare Earth Lanthanide Metals

Actinide Metals

Super Heavy—Radioactive

Chlorine Element

Atomic Structure

Halogens—Halogen chemicals are a special type of element. When they mix with metal, they become a kind of salt. Halogens are super reactive because they like to take an electron from metals. They can be found in column 17 of the element table. Some of them can be found in nature, but most are very dangerous and can hurt you if you touch them. They include fluorine, chlorine, bromine, iodine, and the radioactive elements astatine and tennessine.

Chlorine has had a long and important history. For many years, it was seen as a major breakthrough because it helped solve serious problems in public health and industry. One of its most important early uses was water cleaning. In the early 1900s, people learned that chlorine could kill harmful germs in untreated water. This helped lower the spread of diseases like cholera, typhoid fever, and dysentery. For many communities, this was a big step forward because safer water meant fewer illnesses and fewer deaths. Although some alternatives are in effect, It is still used for cleaning water today.

Chlorine was also used during World War I as a chemical weapon. This was one of the first times a chemical was used in war on a large scale. It caused great suffering and showed how dangerous such weapons could be. Because of this, countries later agreed to limit and ban chemical weapons through international rules and treaties. This helped move the world toward safer and more peaceful ways of dealing with conflict.

Even though chlorine has been very useful in the past and today, many of its old uses are no longer preferred. One reason is that chlorine can create harmful byproducts when it is used in water treatment. Some of these byproducts may be risky to human health and can also hurt fish and other living things in water. Over time, people have become more aware of these problems, so they now look for safer choices.

Another problem is that chlorine can damage pipes and other equipment. When it is used often, it may cause metal parts to wear down faster. This can lead to leaks, repairs, and higher costs. It can also affect the overall quality of water systems. Because of this, many places now want methods that are gentler on their buildings and equipment.

Chlorine is also not as effective as it once was against some germs. A few harmful organisms, such as Cryptosporidium and Giardia (both a diarrheal diseases), can survive chlorine treatment better than other germs. This means chlorine alone may not always give full protection. As a result, water systems often need stronger or more modern methods to make sure water is truly safe.

Because of these concerns, better options have been developed. In water treatment, methods like ultraviolet light, ozone, and membrane filters are now used in many places. These methods can remove or kill many harmful germs without creating as many unwanted byproducts. They can also help protect pipes and other equipment. For this reason, they are becoming more common in modern water systems.

The use of chlorine as a weapon has been strongly restricted. International agreements now ban the making, storing, and using of chemical weapons. These rules support peace and encourage countries to solve problems through discussion instead of violence. This has helped reduce the chance of chlorine being used in such a harmful way again.

(Continued)

Looking to the future, chlorine may still have useful roles in new and positive ways. Researchers are studying chlorine-based materials for clean energy storage. These materials may help improve batteries and other devices that store power from wind and solar energy. If successful, this could support a cleaner and more reliable energy future.

Chlorine may also help in medicine. It is already used in some medicines and cleaning products, but scientists are exploring new ways to use it in treatment. This includes better drug delivery systems, wound care, and safer surgical tools. These ideas could help doctors treat patients more effectively while lowering risks and side effects.

There is also interest in using chlorine in environmental cleanup. Some chlorine-based compounds may help remove pollution from soil and groundwater. They may also help clean wastewater before it returns to rivers or oceans. This could make cleanup work faster and more effective in polluted areas.

In addition, chlorine may help in making new materials. Scientists are studying chlorine compounds that could be used to create lightweight, strong, and long-lasting products. These materials may be useful in construction, transport, and advanced manufacturing. They could lead to better designs and more efficient products in the years ahead.

Chlorine has clearly played many different roles over time. It began as a powerful tool for cleaning water and was later used in ways that raised serious concerns. Today, many of its old uses have been replaced by safer and more effective methods. At the same time, new research is opening the door to fresh and positive uses in energy, medicine, cleanup, and materials. Chlorine's story shows how science can change with time, and how knowledge helps us make better choices for health, safety, and the future.

Uses For Chlorine

Although there are many other products for cleaning, chlorine bleach is still the most commonly used disinfectant and whitening agent for fabrics. It works fast and is cheap, which makes it a popular choice for sanitizing and brightening white clothes. However, it can damage fabric fibers and may harm the environment if used too often. Other common options include hydrogen peroxide, oxygen-based bleach, baking soda, and white vinegar for safer cleaning tasks.

Chlorine is still an important part of cleaning sewage and industrial waste despite the fact that there are other options. It is widely used because it is affordable, easy to apply, and very effective at killing harmful germs. Chlorine is usually added as a liquid or gas. Even though methods like UV light and ozone are available, chlorine is often preferred because it destroys pathogens well and leaves a lasting disinfecting effect.

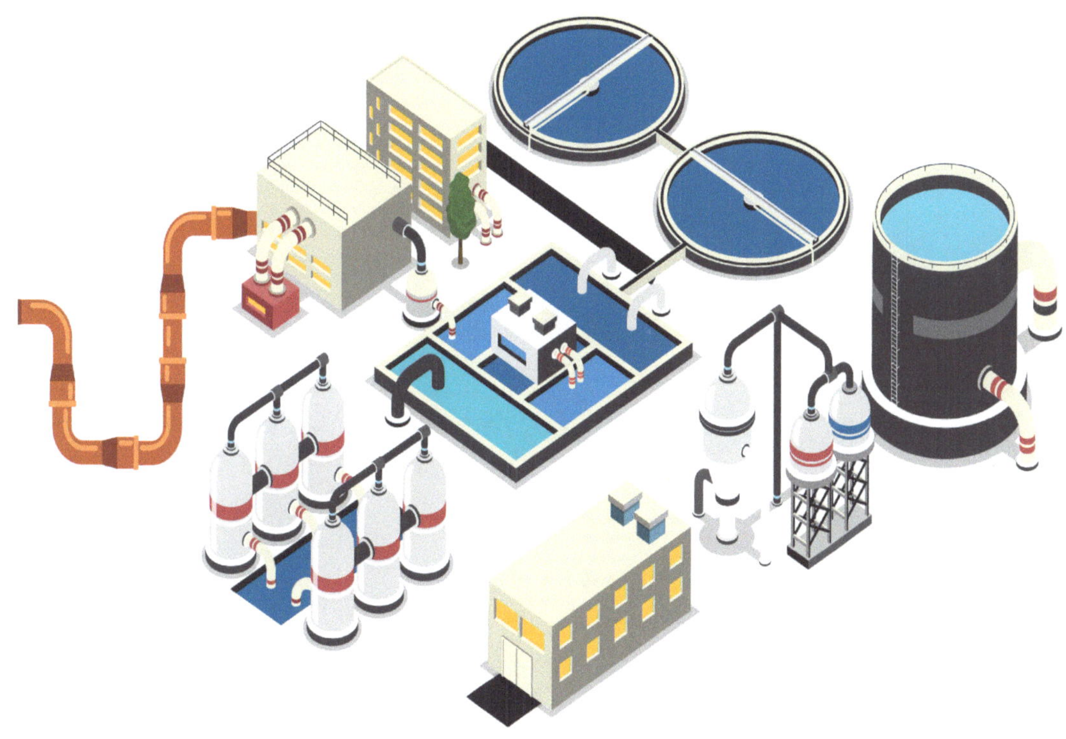

Uses For Chlorine

(Continued)

Many people do not know that chlorine is used in the pulp and paper industry. It helps break down the strong woody cell walls in tree pulp. This makes the pulp softer and easier to turn into paper. The process helps remove color and unwanted materials, too. Without this step, it would be harder to make clean, smooth paper from wood fibers. Chlorine plays an important role in paper production.

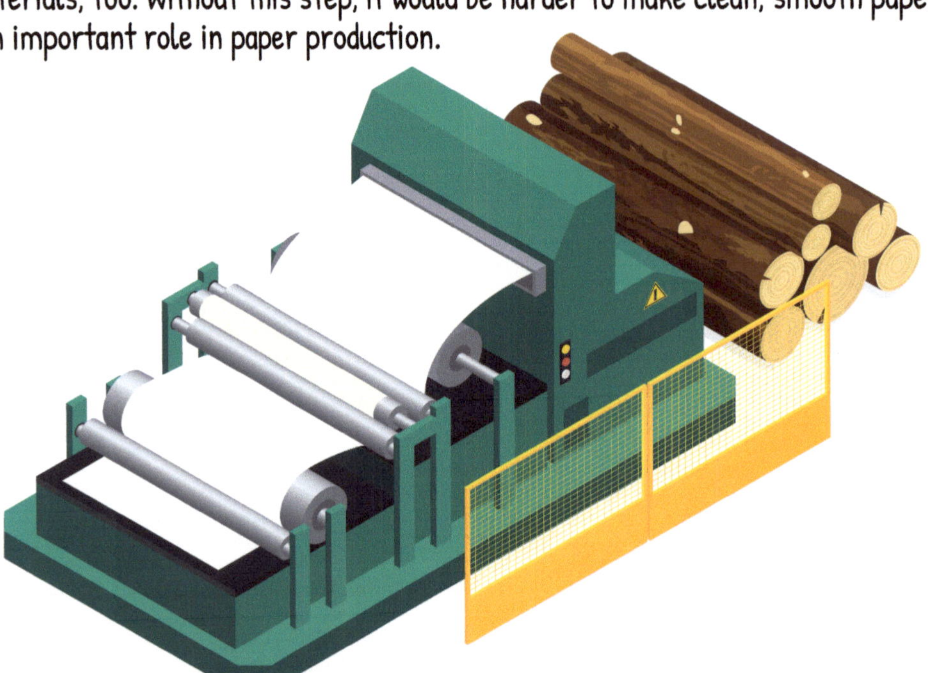

Chlorine is still used in swimming pools because it kills germs like bacteria and viruses by breaking down their cells. However, other pool treatments are also used today. These can lower chlorine use, reduce skin irritation, and cut the strong "chlorine smell" from chloramines. Common options include saltwater systems, ozone, UV light, mineral systems, bromine, and biguanides. Some of these kill germs directly, while others work with a small amount of chlorine for full water protection.

The Source of Chlorine

Chlorine is a chemical element with the symbol "Cl" and atomic number "17". It is widely used in modern life, especially in water treatment, cleaning products, and the manufacture of plastics and other industrial chemicals. In its pure form, chlorine is a greenish-yellow gas with a sharp, irritating smell, so it must be handled carefully. Although most people know chlorine for its role in disinfecting water, it is also a valuable raw material in many manufacturing processes.

Chlorine is not usually found in nature as a free element. Instead, it is obtained from "salt", especially from a mineral called "halite", also known as rock salt. Halite is made mostly of "sodium chloride", the same compound found in table salt and in seawater. Many large salt deposits were formed millions of years ago when ancient seas evaporated and left thick layers of salt behind. These deposits are found in underground mines or close to the Earth's surface in many parts of the world.

To produce chlorine, salt must first be extracted from these deposits. This can be done by "mining" the rock salt directly, using methods such as drilling, cutting, or blasting. In other cases, water is pumped underground to dissolve the salt, and the resulting salty water, called "brine*", is brought back to the surface. The salt is then crushed, cleaned, and prepared for chemical processing.

The Source of Chlorine (Continued)

The main industrial method for making chlorine is "electrolysis". Electrolysis uses electricity to separate a compound into simpler substances. In chlorine production, brine is placed in an electrolytic cell, and an electric current is passed through it. This breaks the sodium chloride solution into different products. "Chlorine gas" is released at one electrode, "hydrogen gas" is produced at the other, and "sodium hydroxide" remains as an important byproduct.

Sodium hydroxide is itself highly useful and is used in making soap, paper, detergents, and many other goods. This means chlorine production also creates valuable chemicals that support other industries. Once chlorine has been produced, it cannot be used immediately. It must first be purified to remove moisture, impurities, and other unwanted substances. After purification, it is often cooled and compressed into a liquid, which makes it easier to store and transport safely.

Liquid chlorine is placed in strong containers or moved through pipelines to factories and treatment plants. There, it is used to produce bleach, plastics, solvents, and a wide range of other chemicals. It is also essential in water treatment, where it helps kill harmful bacteria and other microorganisms, making drinking water safer.

Chlorine production is especially important in countries with large industrial sectors. The United States is one of the world's major producers and users of chlorine, partly because of its salt resources in states such as Louisiana, Texas, and Michigan, as well as its strong chemical industry and transportation systems. Other leading producers include China, India, Germany, and Japan, where large-scale manufacturing and reliable energy supplies support chemical production.

Although chlorine is very useful, its production can have environmental impacts. Mining activities may disturb land, generate waste, and damage habitats if not managed properly. The chemical process can also create pollutants that may affect air and water quality. For this reason, companies must follow strict safety and environmental regulations. Proper handling protects workers, nearby communities, and the environment.

Overall, chlorine plays a major role in modern society. From safe drinking water to industrial manufacturing, it supports many essential services and products that people rely on every day.

Krystix

The Dragon With The Keen Chlorine Tipped Tail

Symbol: Cl Atomic Number: 17 Atomic Mass: 35.45

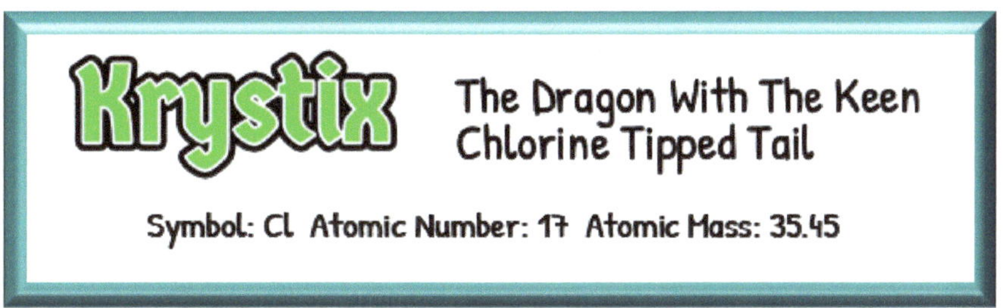

Chlorine resides in Group 17 Period 3 on the Periodic Table.
The Atomic Symbol is Cl. Its Atomic Number is 17. Its Atomic Mass is 35.45.

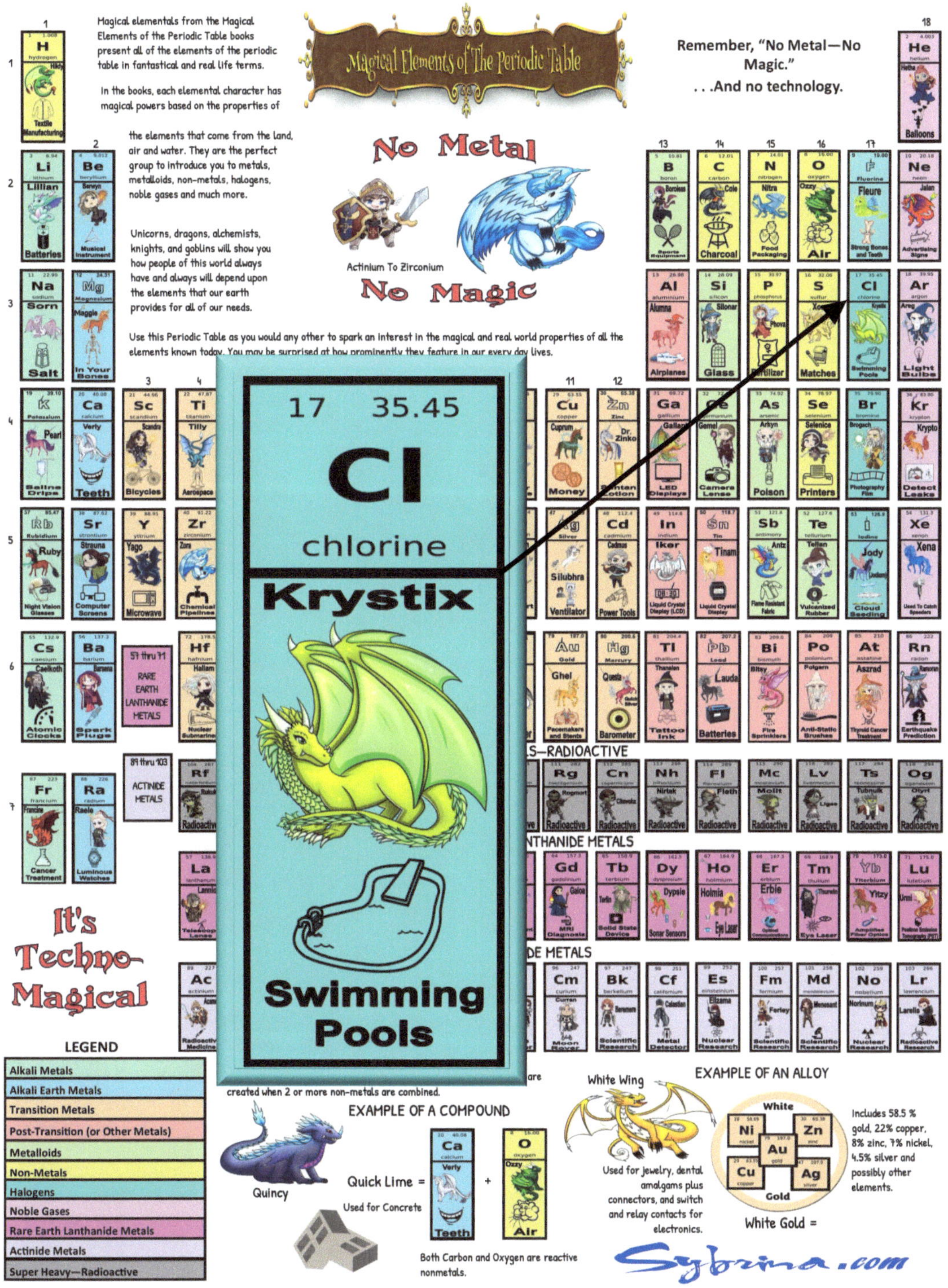

Magical Elements of The Periodic Table

Magical elementals from the Magical Elements of the Periodic Table books present all of the elements of the periodic table in fantastical and real life terms.

In the books, each elemental character has magical powers based on the properties of the elements that come from the land, air and water. They are the perfect group to introduce you to metals, metalloids, non-metals, halogens, noble gases and much more.

Unicorns, dragons, alchemists, knights, and goblins will show you how people of this world always have and always will depend upon the elements that our earth provides for all of our needs.

Use this Periodic Table as you would any other to spark an interest in the magical and real world properties of all the elements known today. You may be surprised at how prominently they feature in our every day lives.

Remember, "No Metal— No Magic." . . .And no technology.

No Metal

No Magic

Actinium To Zirconium

Element	Character	Application
1 1.008 H hydrogen	Hildy	Textile Manufacturing
2 4.003 He helium	Hetha	Balloons
3 6.94 Li lithium	Lillian	Batteries
4 9.012 Be beryllium	Berwyn	Musical Instrument
5 10.81 B boron	Boroleas	Sports Equipment
6 12.01 C carbon	Cole	Charcoal
7 14.01 N nitrogen	Nitra	Food Packaging
8 16.00 O oxygen	Ozzy	Air
9 19.00 F fluorine	Fleure	Strong Bones and Teeth
10 20.18 Ne neon	Jalan	Advertising Signs
11 22.99 Na sodium	Sorn	Salt
12 24.31 Mg magnesium	Maggie	In Your Bones
13 26.98 Al aluminium	Alumna	Airplanes
14 28.09 Si silicon	Silonar	Glass
15 30.97 P phosphorus	Phova	Fertilizer
16 32.06 S sulfur	Xoe	Matches
17 35.45 Cl chlorine	Krystix	Swimming Pools
18 39.95 Ar argon	Areg	Light Bulbs
19 39.10 K Potassium	Pearl	Saline Drips
20 40.08 Ca calcium	Verly	Teeth
21 44.96 Sc scandium	Scandra	Bicycles
22 47.87 Ti titanium	Tilly	Aerospace
23 50.94 V vanadium	Vana	Black Printer Ink
24 52.00 Cr chromium	Crowmist	Stainless Steel
25 54.94 Mn manganese	Mangar	Earth Movers
26 55.85 Fe Iron	Iown	Bicycle Chains
27 58.93 Co cobalt	Coriss	Magnets
28 58.69 Ni nickel	Nix	Guitar Strings
29 63.55 Cu copper	Cuprum	Money
30 65.38 Zn Zinc	Dr. Zinko	Suntan Lotion
31 69.72 Ga gallium	Gallant	LED Displays
32 72.63 Ge germanium	Gemel	Camera Lense
33 74.92 As arsenic	Arkyn	Poison
34 78.97 Se selenium	Selenice	Printers
35 79.90 Br bromine	Brogach	Photography Film
36 83.80 Kr krypton	Krypto	Detect Leaks
37 85.47 Rb Rubidium	Ruby	Night Vision Glasses
38 87.62 Sr strontium	Strauna	Computer Screens
39 88.91 Y yttrium	Yago	Microwave
40 91.22 Zr zirconium	Zora	Chemical Pipelines
41 92.91 Nb niobium	Nonnach	Mag Lev Trains
42 95.95 Mo molybdenum	Maximo	Cutting Tools
43 98 Tc technetium	Tephen	Radio Active Diagnosis
44 101.1 Ru ruthenium	Ruth	Electrical Switches
45 102.9 Rh rhodium	Rovana	Finish for Jewelry
46 106.4 Pd palladium	Paedin	Concert Flute
47 107.9 Ag Silver	Silubhra	Ventilator
48 112.4 Cd cadmium	Cadmus	Power Tools
49 114.8 In indium	Iker	Liquid Crystal Display (LCD)
50 118.7 Sn Tin	Tinam	Liquid Crystal Display
51 121.8 Sb antimony	Antz	Flame Resistant Fabric
52 127.6 Te tellurium	Tellan	Vulcanized Rubber
53 126.9 I Iodine	Jody	Cloud Seeding
54 131.3 Xe xenon	Xena	Used To Catch Speeders

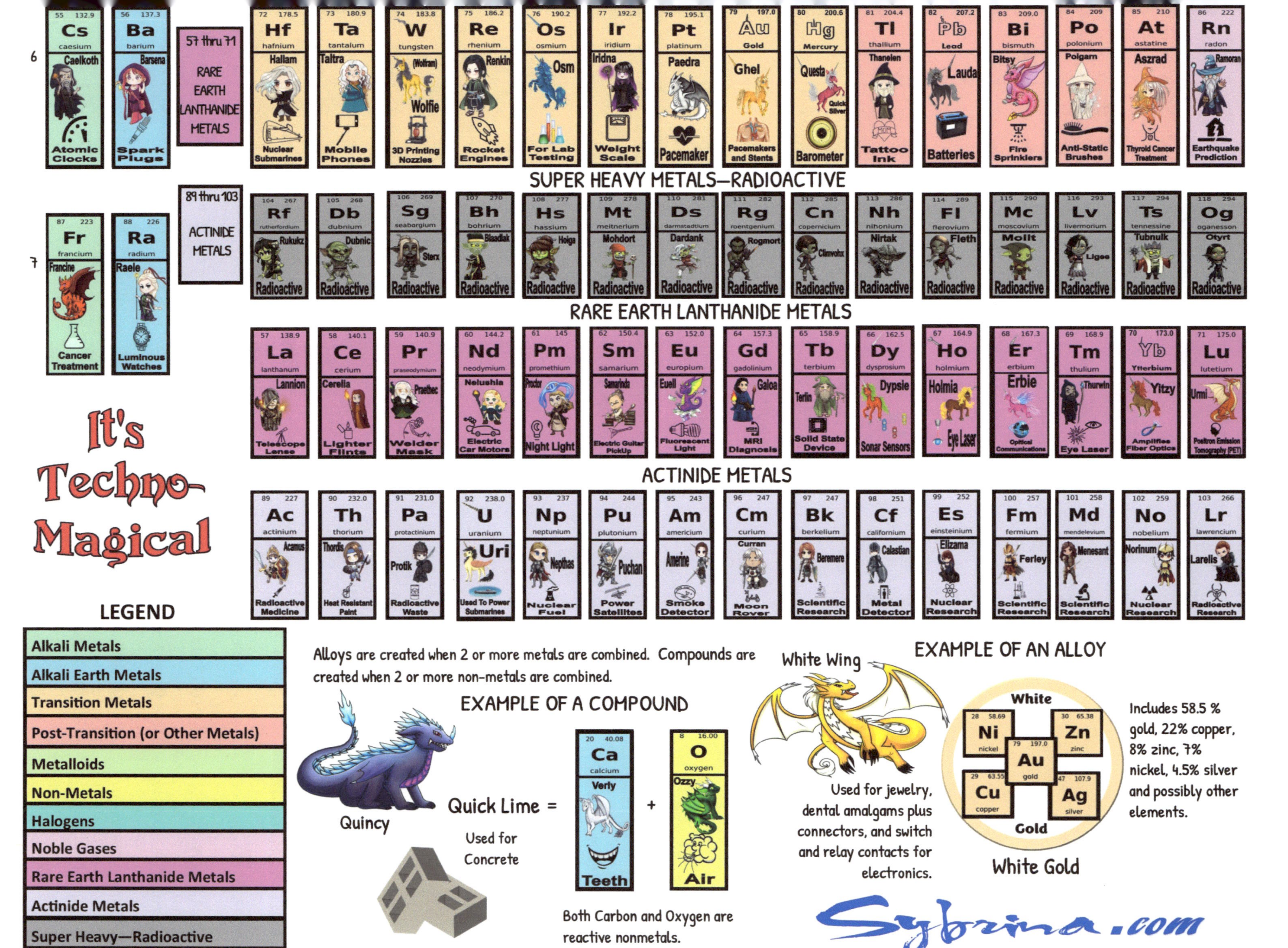

It's Techno-Magical

6 | 55 132.9 Cs caesium Caelkoth — Atomic Clocks | 56 137.3 Ba barium Barsena — Spark Plugs | 57 thru 71 RARE EARTH LANTHANIDE METALS | 72 178.5 Hf hafnium Hallam — Nuclear Submarines | 73 180.9 Ta tantalum Taltra — Mobile Phones | 74 183.8 W tungsten (Wolfram) Wolfie — 3D Printing Nozzles | 75 186.2 Re rhenium Renkin — Rocket Engines | 76 190.2 Os osmium Osm — For Lab Testing | 77 192.2 Ir iridium Iridna — Weight Scale | 78 195.1 Pt platinum Paedra — Pacemaker | 79 197.0 Au gold Ghel — Pacemakers and Stents | 80 200.6 Hg mercury Questa Quick Silver — Barometer | 81 204.4 Tl thallium Thanelen — Tattoo Ink | 82 207.2 Pb lead Lauda — Batteries | 83 209.0 Bi bismuth Bitsy — Fire Sprinklers | 84 209 Po polonium Polgarn — Anti-Static Brushes | 85 210 At astatine Aszrad — Thyroid Cancer Treatment | 86 222 Rn radon Ramoran — Earthquake Prediction

SUPER HEAVY METALS—RADIOACTIVE

7 | 87 223 Fr francium Francine — Cancer Treatment | 88 226 Ra radium Raele — Luminous Watches | 89 thru 103 ACTINIDE METALS | 104 267 Rf rutherfordium Rukukz — Radioactive | 105 268 Db dubnium Dubnic — Radioactive | 106 269 Sg seaborgium Sterx — Radioactive | 107 270 Bh bohrium Blaadlak — Radioactive | 108 277 Hs hassium Hoiga — Radioactive | 109 278 Mt meitnerium Mohdort — Radioactive | 110 281 Ds darmstadtium Dardank — Radioactive | 111 282 Rg roentgenium Rogmort — Radioactive | 112 285 Cn copernicium Climvotx — Radioactive | 113 286 Nh nihonium Nirtak — Radioactive | 114 289 Fl flerovium Fleth — Radioactive | 115 290 Mc moscovium Mollt — Radioactive | 116 293 Lv livermorium Ligee — Radioactive | 117 294 Ts tennessine Tubnulk — Radioactive | 118 294 Og oganesson Otyrt — Radioactive

RARE EARTH LANTHANIDE METALS

57 138.9 La lanthanum Lannion — Telescope Lense | 58 140.1 Ce cerium Cerelia — Lighter Flints | 59 140.9 Pr praseodymium Praethec — Welder Mask | 60 144.2 Nd neodymium Nelushia — Electric Car Motors | 61 145 Pm promethium Proctor — Night Light | 62 150.4 Sm samarium Samarinda — Electric Guitar PickUp | 63 152.0 Eu europium Euell — Fluorescent Light | 64 157.3 Gd gadolinium Galoa — MRI Diagnosis | 65 158.9 Tb terbium Terfin — Solid State Device | 66 162.5 Dy dysprosium Dypsie — Sonar Sensors | 67 164.9 Ho holmium Holmia — Eye Laser | 68 167.3 Er erbium Erbie — Optical Communications | 69 168.9 Tm thulium Thurwin — Eye Laser | 70 173.0 Yb Ytterbium Yitzy — Amplifies Fiber Optics | 71 175.0 Lu lutetium Urmi — Positron Emission Tomography (PET)

ACTINIDE METALS

89 227 Ac actinium Acamus — Radioactive Medicine | 90 232.0 Th thorium Thordis — Heat Resistant Paint | 91 231.0 Pa protactinium Protik — Radioactive Waste | 92 238.0 U uranium Uri — Used To Power Submarines | 93 237 Np neptunium Nephas — Nuclear Fuel | 94 244 Pu plutonium Puchan — Power Satellites | 95 243 Am americium Amerine — Smoke Detector | 96 247 Cm curium Curran — Moon Rover | 97 247 Bk berkelium Beremere — Scientific Research | 98 251 Cf californium Calastian — Metal Detector | 99 252 Es einsteinium Elizama — Nuclear Research | 100 257 Fm fermium Ferley — Scientific Research | 101 258 Md mendelevium Menesant — Scientific Research | 102 259 No nobelium Norinum — Nuclear Research | 103 266 Lr lawrencium Larelis — Radioactive Research

LEGEND
Alkali Metals
Alkali Earth Metals
Transition Metals
Post-Transition (or Other Metals)
Metalloids
Non-Metals
Halogens
Noble Gases
Rare Earth Lanthanide Metals
Actinide Metals
Super Heavy—Radioactive

Alloys are created when 2 or more metals are combined. Compounds are created when 2 or more non-metals are combined.

EXAMPLE OF A COMPOUND
Quincy
Quick Lime =
Used for Concrete
20 40.08 Ca calcium Verly — Teeth
+
8 16.00 O oxygen Ozzy — Air
Both Carbon and Oxygen are reactive nonmetals.

EXAMPLE OF AN ALLOY
White Wing
White
28 58.69 Ni nickel
30 65.38 Zn zinc
79 197.0 Au gold
29 63.55 Cu copper
47 107.9 Ag silver
Gold
White Gold
Includes 58.5 % gold, 22% copper, 8% zinc, 7% nickel, 4.5% silver and possibly other elements.
Used for jewelry, dental amalgams plus connectors, and switch and relay contacts for electronics.

Sybrina.com

All Of The Periodic Table Elements Listed Alphabetically

Element Listed In Red Is Featured In This Book

ACTINIUM—*AC*—89

ALUMINUM—*AL*—13

AMERICIUM—*AM*—95

ANTIMONY—*SB*—51

ARGON—*AR*—18

ARSENIC—*AS*—33

ASTATINE—*AT*—85

BARIUM—*BA*—56

BERKELIUM—*BK*— 97

BERYLLIUM—*BE*—4

BISMUTH—*BI*—83

BOHRIUM—*BH*—107

BORON—*B*—5

BROMINE—*BR*—35

CADMIUM—*CD*—48

CALCIUM (Vital)—*CA*—20

CALIFORNIUM—*CF*—98

CARBON—*C*—6

CERIUM—*CE*—58

CESIUM—*CS*—55

CHLORINE (Keen)—*CL*—17

CHROMIUM—*CR*—24

COBALT—*CO*—27

COPERNICIUM—*CN*—112

COPPER—*CU*—29

CURIUM—*CM*—96

DARMSTADTIUM—*DS*—110

DUBNIUM—*DB*—105

DYSPROSIUM—*DY*—66

ERBIUM—*ER*—68

EINSTEINIUM—*ES*—99

EUROPIUM—*EU*—63

FERMIUM—*FM*—100

FLEROVIUM—*FL*—114

FLUORINE—*F*—9

FRANCIUM—*FR*—87

GADOLINIUM—*GD*—64

GALLIUM—*GA*—31

GERMANIUM—*GE*—32

GOLD—*AU*—79

HAFNIUM—*HF*—72

HASSIUM—*HS*—108

HELIUM—*HE*—2

HOLMIUM—*HO*—67

HYDROGEN—*H*—1

INDIUM—*IN*—49

IODINE *(JODIUM)*—*I*—53

IRIDIUM—*IR*—77

IRON—*FE*—26

KRYPTON—*KR*—36

LANTHANUM—*LA*—57

LAWRENCIUM—*LR*—103

LEAD—*PB*—82

LITHIUM—*LI*—3

LIVERMORIUM—*LV*—116

LUTETIUM (Unique)—*LU*—71

MAGNESIUM—*MG*—12

MANGANESE—*MN*—25

MEITNERIUM—*MT*—109

MENDELEVIUM—*MD*—101

MERCURY *(QUICK SILVER)*—*HG*—80

MOLYBDENUM—*MO*—42

MOSCOVIUM—*MC*—115

NEODYMIUM—*ND*—60

NEON (Jazzy)—*NE*—10

NEPTUNIUM—*NP*—93

NICKEL—*NI*—28

NIHONIUM—*NH*—113

NIOBIUM—*NB*—41

NITROGEN—*N*—7

NOBELIUM—*NO*—102

OGANESSON—*OG*—118

OSMIUM—*OS*—76

OXYGEN—*O*—8

PALLADIUM—*PD*—46

PHOSPHORUS—*P*—15

PLATINUM—*PT*—78

PLUTONIUM—*PU*—94

POLONIUM—*PO*—84

POTASSIUM—*K*—19

PRASEODYMIUM—*PR*—59

PROMETHIUM—*PM*—61

PROTACTINIUM—*PA*—91

RADIUM—*RA*—88

RADON—*RN*—86

RHENIUM—*RE*—75

RHODIUM—*RH*—45

ROENTGENIUM—*RG*—111

RUBIDIUM—*RB*—37

RUTHENIUM—*RU*—44

RUTHERFORDIUM—*RF*—104

SAMARIUM—*SM*—62

SCANDIUM—*SC*—21

SEABORGIUM—*SG*—106

SELENIUM—*SE*—34

SILICON—*SI*—14

SILVER—*AG*—47

SODIUM—*NA*—11

STRONTIUM—*SR*—38

SULFUR (Xanthous)—*S*—16

TANTALUM—*TA*—73

TECHNETIUM—*TC*—43

TELLURIUM—*TE*—52

TENNESSINE—*TS*—117

TERBIUM—*TB*—65

THALLIUM—*TI*—81

THORIUM—*TH*—90

THULIUM—*TM*—69

TIN—*SN*—50

TITANIUM—*TI*—22

TUNGSTEN—*W (WOLFRAM)*—74

URANIUM—*U*—92

VANADIUM—*V*—23

XENON—*XE*—54

YTTERBIUM—*YB*—70

YTTRIUM—*Y*—39

ZINC—*ZN*—30

ZIRCONIUM—*ZR*—40

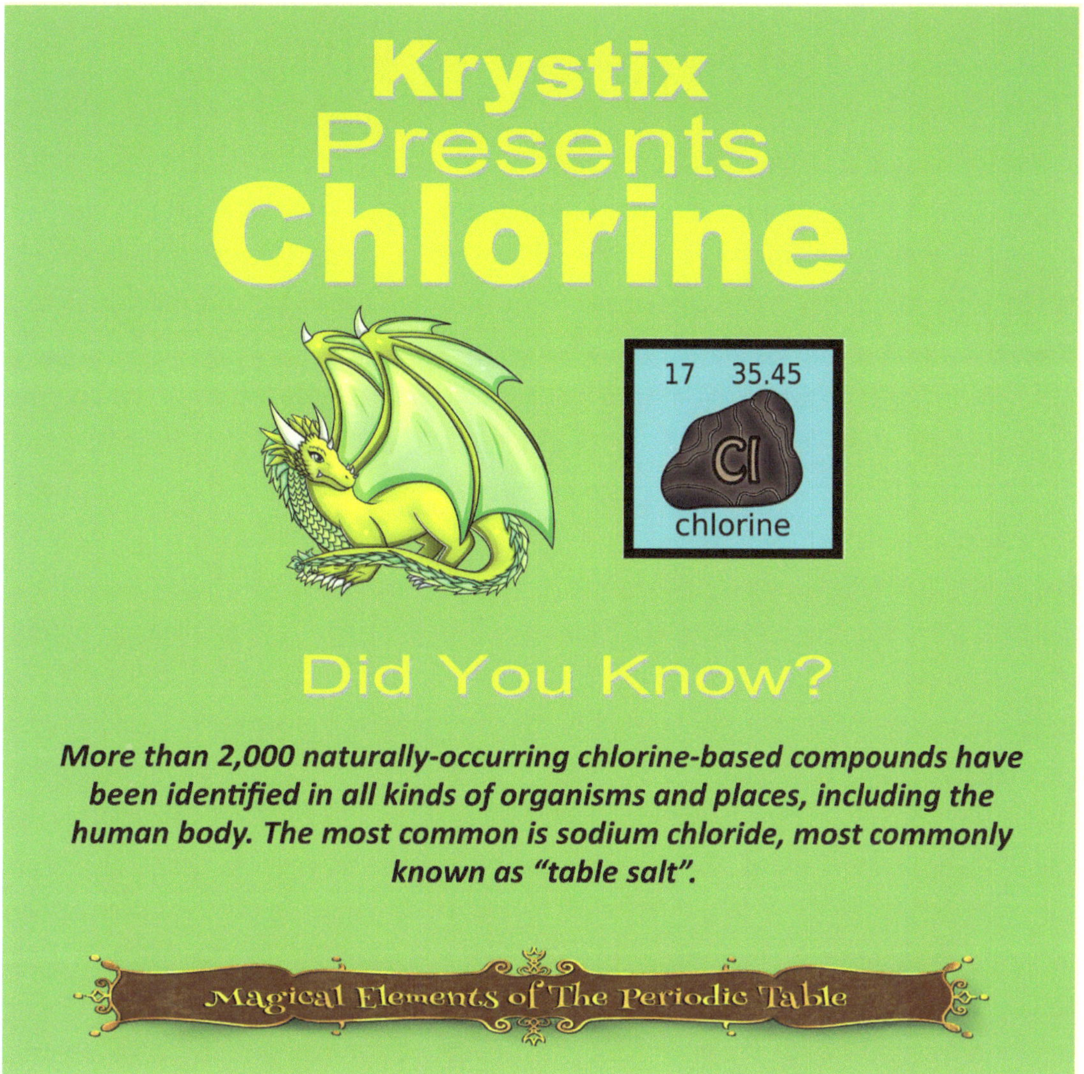

- Chlorine was first made and seen as a separate gas in 1774 by the Swedish chemist Carl Wilhelm Scheele. He thought it was a gas containing oxygen. Earlier people, such as Jan Baptist van Helmont in the 1630s, may have seen similar gases, but Scheele studied it carefully. He called it "dephlogisticated muriatic acid air." In 1810, the English chemist Humphry Davy showed that chlorine was really an element. He proved it could not be broken into simpler parts, so it was not a compound but a true element.

- In 1810, English chemist Sir Humphry Davy gave the element chlorine its name. He based it on the Greek word "khloros" (χλωρός), which means "pale green," "greenish-yellow," or "yellowish-green." Davy picked this name because it matched the gas's color. Chlorine is known for its yellow-green look, which helped make the name easy to remember.

- Chlorination is one of many methods that can be used to disinfect water. This method was first used over a century ago, and is still used today to combat waterborne diseases. The use of Chlorine has proven effective against bacteria and viruses and makes our water safer.

- Chloride is an ion of chlorine. It occurs when a chlorine atom gains one electron which forms a negatively charged ion. Despite its toxicity, the chlorine ion, chloride, is the ninth most abundant element in the human body. It is essential for human nerve function because it regulates the excitability of neurons and enables efficient nerve impulse transmission. Most chloride in our diets come from salt.

Did You Know? (continued)

- Volcanoes mostly release chlorine as hydrogen chloride gas, not as large amounts of elemental chlorine gas. Hydrogen chloride is the main chlorine gas from volcanic activity. In the air, it can react with oxygen and other gases to make small amounts of reactive chlorine compounds. This can happen when hot volcanic gas mixes with air or when lava touches seawater. These reactions are important because they can affect the air near a volcano and may help create other chemical changes.
- If all the chlorine in Earth's oceans were turned into gas and released into the air, it would be very heavy. In fact, it would weigh about five times more than the atmosphere we have now. That means the amount of chlorine in the oceans is huge. The air would become much thicker and heavier than it is today. This shows how much material is hidden in the oceans.
- Chlorine was the first chemical weapon used in World War I. German forces first used it in 1915. At room temperature, chlorine is a very poisonous gas. It is also about 2.5 times heavier than air. Because of this, it sank down into trenches instead of rising away. This made it especially deadly for soldiers hiding there. The gas caused serious harm and fear on the battlefield. Its use marked a dangerous new kind of warfare. Chlorine showed how chemical weapons could be used to attack enemy troops in a cruel and hidden way.
- Chlorine is a chemical that is found in common table salt, also called sodium chloride. Salt is used every day in many homes to add flavor to food. The chlorine in salt is not the same as the dangerous chlorine gas used for cleaning or in swimming pools. In salt, chlorine is joined with sodium to make a stable compound. This compound is safe to eat in small amounts and is an important part of our diet. So, chlorine is part of something we use often, but in a form that is very different from pure chlorine.
- This element reacts very quickly. It can even start fire on its own. It may set steel wool, paper, and some metals burning without any spark or flame. Because of this, it must be handled very carefully.
- The "chlorine smell" in swimming pools is not the smell of chlorine alone. It comes mostly from chloramines. These are made when chlorine mixes with sweat, urine, and other body waste in the water. So, a strong pool smell can actually mean the pool needs better cleaning or more fresh chlorine. Proper pool care helps keep the water safer and the air around it fresher.

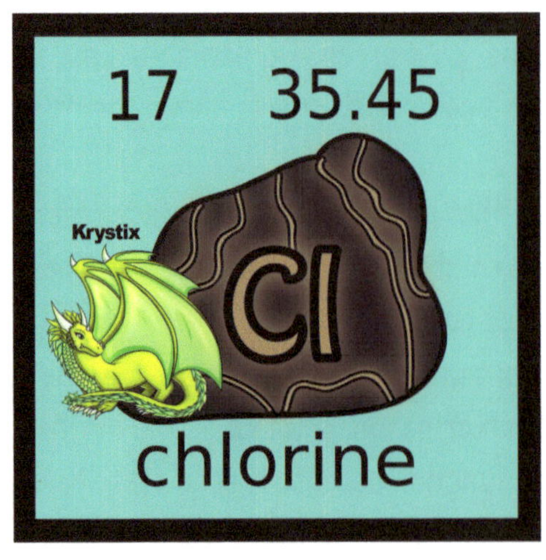

Structure Of Elements In The Periodic Table

Periodic tables are laid out in rows and columns.

Each element is placed in a specific location because of its atomic structure. Elements are arranged in Families.

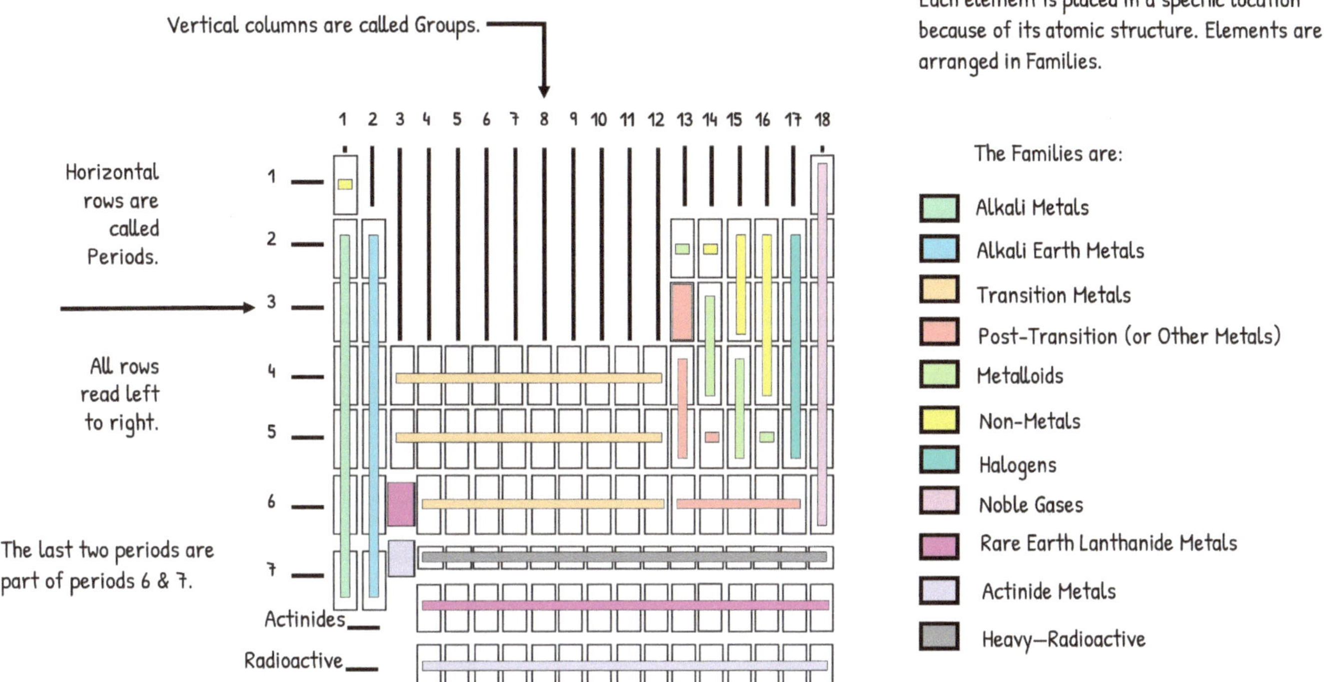

The term 'Element' is used to describe atoms with specific characteristics.
Every element in the first column or Group has 1 electron in the outer orbital (shell).
Every element in the second column (group two) has two electrons in the element's outer orbital.
The number designation of each Group represents the number of electrons in the element's outer orbital—
except for Group 18, Period 1—Helium. It only has 2 electrons.
Those electrons, called Valence Electrons, are what chemically bond with other elements.

Atomic Structure of Element: The atomic structure of an element refers to the arrangement of protons and neutrons in the nucleus of the atom, and the electrons in the electron cloud around the nucleus. Group 1, Period 1—Hydrogen is the only element that has no neutrons.

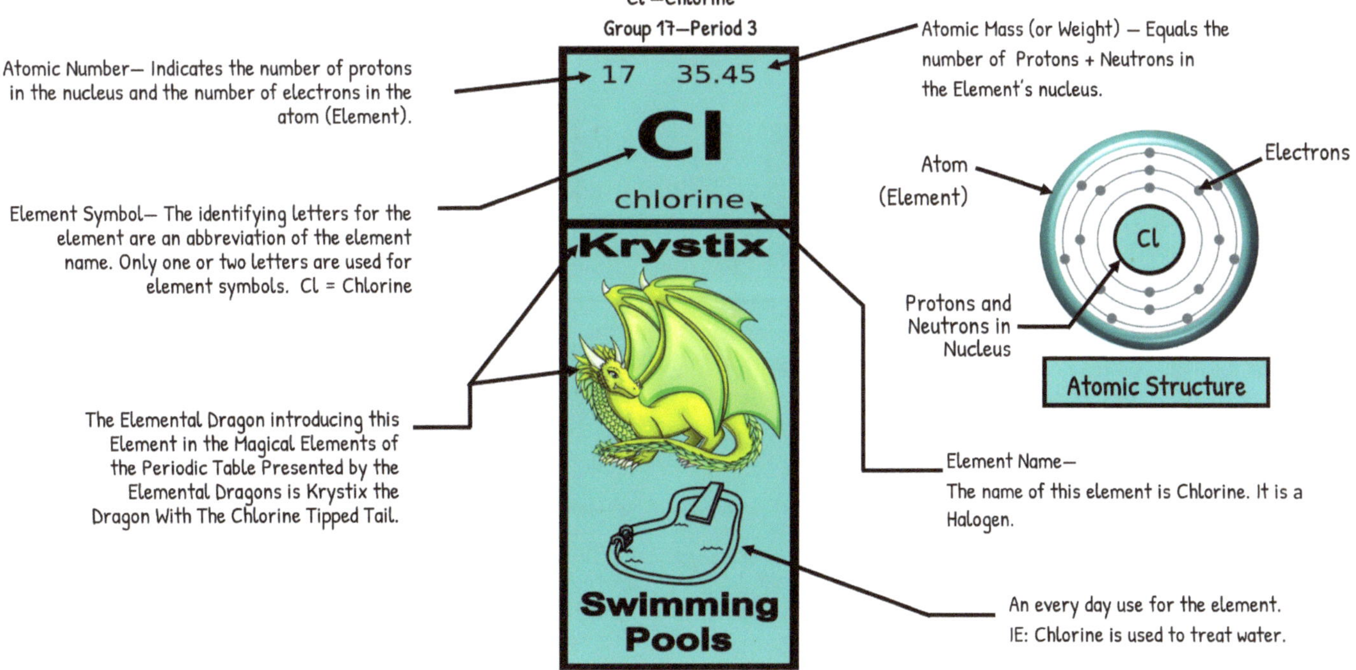

Types of Elements On The Periodic Table

Alkali Metals—Some metals on the periodic table are soft and shiny. They are so soft that they can be cut with a knife! These metals are excited to give away electrons to elements in need, making them highly reactive. Ther electron transfer creates a compound known as a salt. Surprisingly, these metals are not found in nature alone; they must be extracted from other sources. Examples of these metals include lithium, sodium, potassium, rubidium, cesium, and francium.

Alkali Earth Metals—The elements in column 2 of the periodic table have 2 outer electrons in their shell. Ther makes them very active with nonmetals that need electrons to stay stable. When they react, they make something called a salt. They are often found in nature all by themselves, and they can even conduct electricity. The elements are beryllium, magnesium, calcium, strontium, barium, and radium.

Post-Transition (or other Metals)— Elements directly to the right of the transition metals. They are known as "poor metals: and are soft and brittle. These include aluminum, gallium, indium, tin, thallium, lead, bismuth, zinc, cadmium and mercury.

Transition Metal—The main metals are found in the middle and bottom rows of the periodic table. They look like metal, can conduct electricity, can bend and be shaped easily. The period 4 transition metals are scandium, titanium, vanadium, chromium, manganese, iron, cobalt, nickel, copper, and zinc. The period 5 transition metals are yttrium, zirconium, niobium, molybdenum, technetium, ruthenium, rhodium, palladium, silver, and cadmium. The period 6 transition metals are lanthanum, hafnium, tantalum, tungsten, rhenium, osmium, iridium, platinum, gold, and mercury. The period 7 transition metals are the naturally-occurring actinium, and the artificially produced elements rutherfordium, dubnium, seaborgium, bohrium, hassium, meitnerium, darmstadtium, and roentgenium.

Metalloids—The elements called metalloids are a mix of metals and nonmetals. They look like metals, but can't conduct electricity very well. They also break easily and act like nonmetals. These include boron, silicon, germanium, arsenic, antimony, tellurium, astatine, and polonium.

Non-Metals—These elements reside in columns 15-17, and can be gases, liquids, or solids. They don't conduct heat or electricity. The solids are brittle, and they have no metallic luster. They readily accept electrons from metals to form salts. These include nitrogen, oxygen, fluorine, chlorine, bromine, and iodine.

Halogens—Halogen chemicals are a special type of element. When they mix with metal, they become a kind of salt. Halogens are super reactive because they like to take an electron from metals. They can be found in column 17 of the element table. Some of them can be found in nature, but most are very dangerous and can hurt you if you touch them. They include fluorine, chlorine, bromine, iodine, and the radioactive elements astatine and tennessine.

Noble Gases—These elements reside in column 8. They are all odorless, colorless gases that are chemically very stable (inert). They don't generally form compounds by bonding with another element. These include helium, neon, argon, krypton, xenon, and radon.

Lanthanide Rare Earth Minerals—The Japanese call them "the seeds of technology." The US Department of Energy calls them "technology metals." These elements have atomic numbers 57-71. They are vital to industry. They can be added to metals to strengthen them to make alloys such as stainless steel, used to refine crude oil, and are crucial in producing technology—electronics, telecommunications, and metal devices to name a few. They are lanthanum, cerium, praseodymium, neodymium, promethium, samarium, europium, gadolinium, terbium, dysprosium, holmium, erbium, thulium,

Actinide Metals—Any of a series of chemically similar metallic elements with atomic numbers ranging from 89 (actinium) to 103 (lawrencium). All of these elements are radioactive, and two of the elements, uranium and plutonium, are used to generate nuclear energy. The lanthanides and actinides are sometimes called the inner transition metals, referring to their properties and position on the table. They are actinium, thorium, protactinium, uranium, neptunium, plutonium,

Super Heavy—Radioactive—Superheavy elements are those elements with a large number of protons in their nucleus. Elements with more than 92 protons are unstable; they decay to lighter nuclei with a characteristic half-life. They do not occur in large quantities (if at all) naturally on earth, and only exist briefly under highly controlled circumstances. They include lawrencium, rutherfordium, dubnium, seaborgium, bohrium, hassium, meitnerium, darmstadtium, roentgenium, copernicium, nihonium, flerovium, moscovium, livermorium, tennessine, and oganesson.

Alloys

An **alloy** is a mixture of chemical elements of which at least one is a metal. An alloy is a solid. Unlike chemical compounds with metallic bases, an alloy will retain all the properties of a metal in the resulting material, such as electrical conductivity, ductility, opacity, and luster, but may have properties that differ from those of the pure metals, such as increased strength or hardness. In some cases, an alloy may reduce the overall cost of the material while preserving important properties. In other cases, the mixture imparts synergistic properties to the constituent metal elements such as corrosion resistance or mechanical strength. Some of the most common alloys are

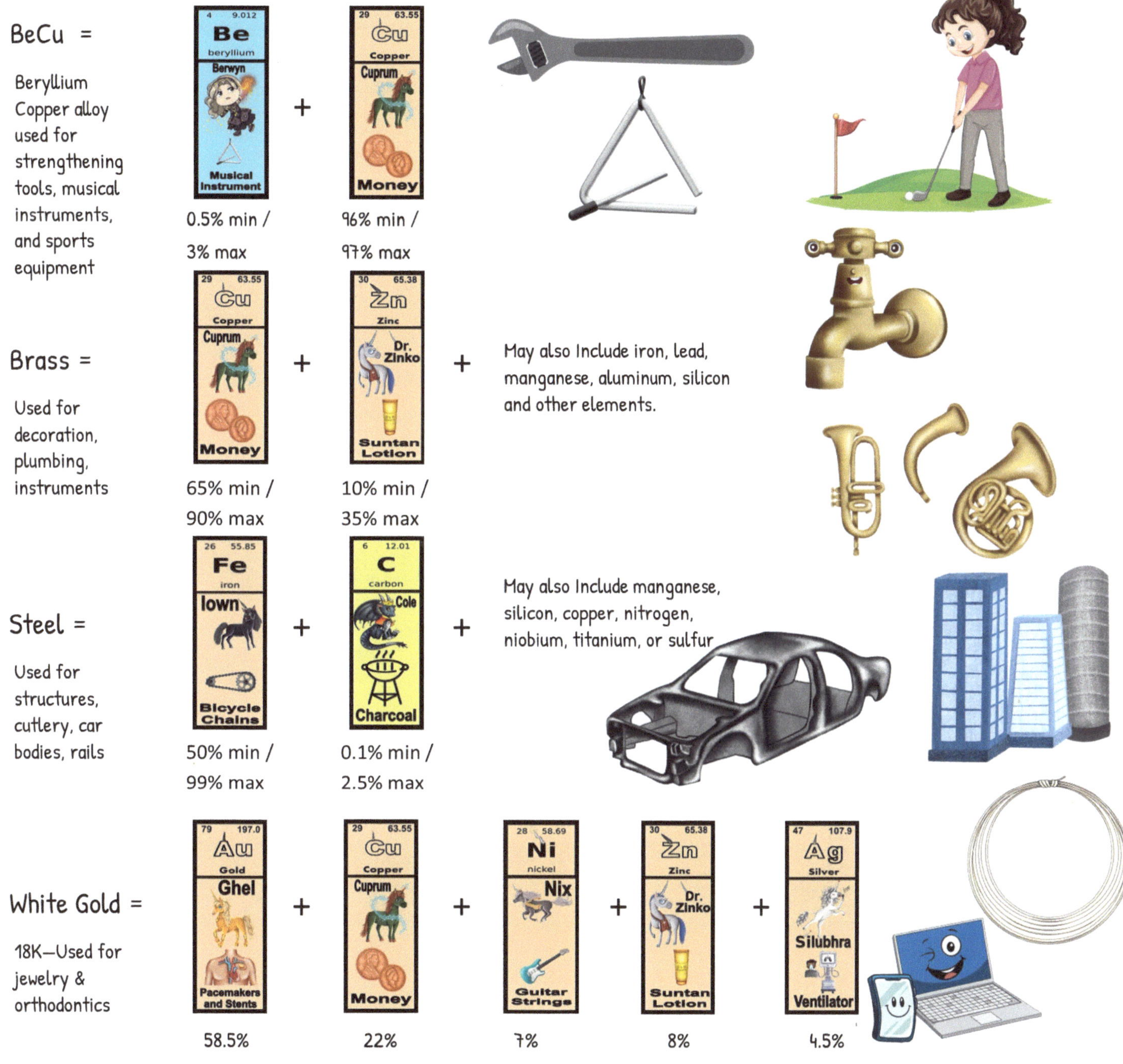

Some other common alloys are Bronze, Cast Iron, Cupronickel, Magnalium, Mischmetal, Nichrome, Nitinol, Pewter, Solder, Sterling Silver and Tungsten Carbide.

The above chart only shows a few of the hundreds of metal combinations. For instance, 24 carat gold is a pure naturally occurring yellow metal. There are four basic shades of gold alloys: yellow gold, white gold, rose gold, and green gold. A huge range of other colored golds are also possible, including red (gold and copper), grey (gold, iron and copper), purple (gold and aluminum), blue (gold and iron) and black (gold and cobalt), depending on the amounts of different metals alloyed together.

Compounds

A **compound** is a substance formed due to the chemical union (a chemical reaction) between two or more atoms or molecules. Ideally there should be two or more elements to form a compound. Most of the compounds are from a non-metallic origin.

Fluorocarbon =

Used for— waterproofing agents, lubricants, sealants, and leather conditioners.

+

Both Carbon and Fluorine are reactive nonmetals.

Sodium Fluoride =

Used for—the fluoridation of drinking water, in toothpaste, in metallurgy, and as a flux, and is also used in pesticides and rat poison.

+

Sodium Fluoride is a simple ionic compound, made of the sodium (Na+) cation and fluoride (F-), an anion of Flourine.

Potassium Iodide =

It's medical use is to block absorption of radioactive iodine by the thyroid gland.

+

Potassium iodide (also called KI) is a salt of stable (not radioactive) iodine. Potassium is an alkali metal and Iodine is a reactive nonmetal.

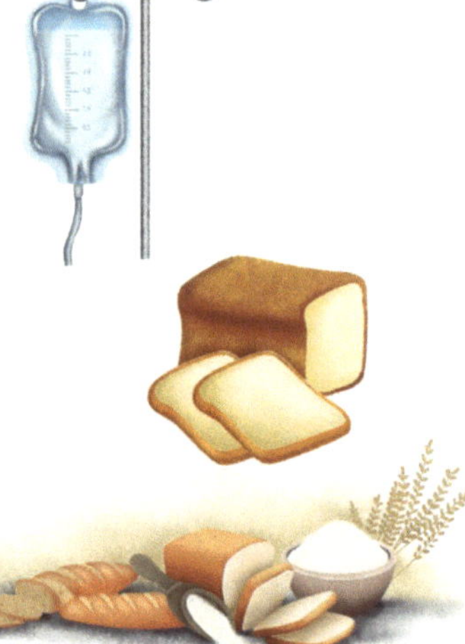

Calcium Bromate =

It is used as a bread dough and flour "improver" or conditioner

+

Calcium is an alkaline earth metal and Bromine is a reactive nonmetal.

Can you guess the most important compound of all?

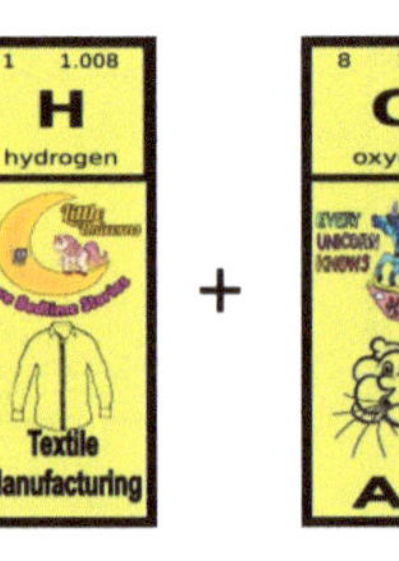

+

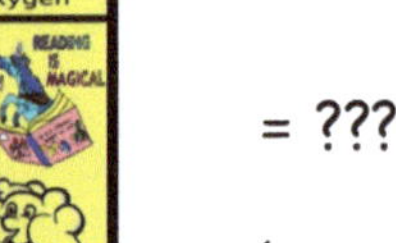

= ??????

(Answer can be found below.)

The above chart only shows a few of the millions of combinations of elements that create compounds.

Iodine, alone has 37 compounds. Some of them are Ammonium iodide (NH4I), Cesium iodide (CsI), Copper(I) iodide (CuI), Hydroiodic acid (HI), Iodic acid (HIO3), Iodine cyanide (ICN), Iodine heptafluoride (IF7), Iodine pentafluoride (IF5), Lead(II) iodide (PbI2), Lithium iodide (LiI), Nitrogen triiodide (NI3), Potassium iodate (KIO3), Potassium iodide (KI), Sodium iodate (NaIO3), and Sodium iodide (NaI).

Hydrogen is used in more compounds (nearly 100) than any other element because it can form bonds with almost all metals, metalloids, and non-metals. Some of the most common are water (H_2O), Hydrogen Peroxide (H_2O_2) and table sugar ($C_{12}H_{22}O_{11}$).

Neon forms no known compounds.

Definitions

Atomic Structure of Element: The atomic structure of an element refers to the arrangement of protons and neutrons in the nucleus of the atom, and the electrons in the electron cloud around the nucleus.

Atomic Number: An element's atomic number refers to the number of protons it has in its nucleus. In a neutral atom the number of protons always equals the number of electrons.

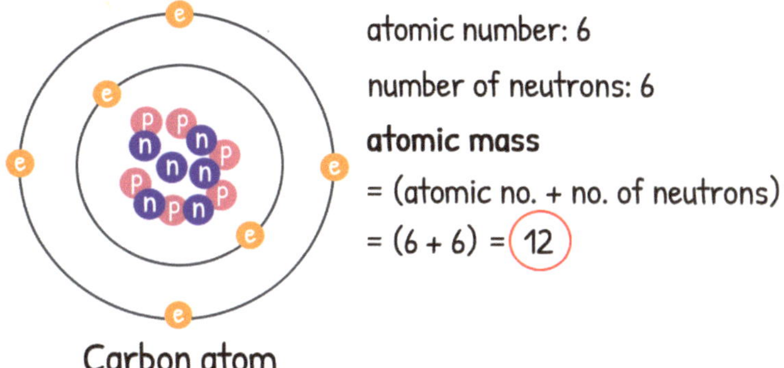

Atomic Weight (Mass) of Element: The atomic mass of an element is how heavy it is. It is made up of protons and neutrons that are in the middle of the element. Some elements have different versions with different amounts of neutrons, but they still have the same amount of protons. The atomic mass is the average weight of all these versions of the element.

Allotrope: Allotropes are different forms of an element that look and act different, but are made of the same stuff. Some elements have more than one form. For instance, carbon can be a shiny diamond or a gray pencil lead called graphite.

Isotope: Isotopes are different types of atoms that have the same parts, like protons and electrons, but they have a different number of neutrons. For example, the three most stable isotopes of hydrogen: protium (A = 1), deuterium (A = 2), and tritium (A = 3).

Crystalline Structure of Element: The crystalline structure of an element is how its atoms, ions, or molecules stick together in a pattern to make a cool crystal shape.

Ferrous and Non-Ferrous Metals: When we say ferrous metal, it means that iron is a big part of the metal. But if there's only a little bit of iron in the metal, we call it non-ferrous. The word "ferrous" comes from Latin and means iron, which is why iron's symbol is Fe.

Ductile Metals: These are capable of being made into long, thin wire or thread. Copper and Silver are ductile metals.

Malleable Metals: These can be hammered or rolled into thin sheets without cracking or breaking. Gold is malleable.

Ferromagnetic: Materials that are strongly attracted to a magnet. Such materials can be permanently magnetized. These include the elements iron, nickel and cobalt and their alloys, some alloys of rare-earth metals, and some naturally occurring minerals such as lodestone.

Magnetostriction—Ther is the term for a special thing that happens to magnetic materials. When these materials get turned into magnets, they also change their shape or size.

Paramagnetic: Slightly attracted to a magnetic field, but do not retain magnetic properties once the field is removed.

Diamagnetic: Slightly repelled by a magnetic field, but do not retain magnetic properties once the field is removed.

Electrical Properties: Conductor—a thing that lets electricity flow through it. Semi-conductor—a special material (usually silicon) that can conduct electricity, but not as well as metal. Insulator (non-conductor)—a material (usually glass) that stops electricity from flowing.

Reactive Gas: These gases are really good at reacting with stuff! They are called "sticky gases" because they can react to things like plastic and wet surfaces when they touch them. These are nitrogen, oxygen, hydrogen, carbon dioxide, fluorine, and chlorine.

Non-Reactive Gas: An inert gas is like a super shy gas that doesn't like to hang out with other chemicals. It doesn't make any new friends by reacting with them, so it doesn't form any chemical compounds. We also call these special gases "noble gases."

What Makes Chlorine Seem Magical?

There is something almost magical about chlorine. It is found in the periodic table as element 17, but it is more than just a science fact. Chlorine has a strange power that makes it seem special and mysterious.

At room temperature, chlorine is a pale green gas. It has a strong, sharp smell that can make it hard to breathe. This gas is very reactive, which means it joins easily with other elements to form new substances. Chlorine does this because it wants to gain one electron and become more stable. This strong reaction makes chlorine feel exciting and a little dangerous at the same time.

One reason chlorine seems magical is the many ways it helps people every day. The most important use is in cleaning water. Chlorine can kill bacteria, viruses, and other tiny germs that can make people sick. Water treatment plants use it all over the world to make drinking water safer. Because of this, chlorine has helped save many lives. It is often linked with cleanliness and safety.

Chlorine is also used in many industries. It helps make plastics, solvents, and many common products. It is part of bleach, which people use to clean and whiten clothes and surfaces. Since chlorine is found in so many things we use at home, it can seem almost like a hidden force that helps keep our world bright and clean.

The history of chlorine is also interesting. A Swedish chemist named Carl Wilhelm Scheele discovered it in 1774. At first, people knew it mostly for its ability to bleach things. Later, in the 1800s, people began to understand its value in medicine and public health. During the American Civil War, chlorine-based cleaning substances helped lower death rates among wounded soldiers. This showed that chlorine could fight invisible dangers and protect human life.

But chlorine also has a darker side. In World War I, chlorine gas was used as a weapon. Because it is poisonous and heavier than air, it could cause terrible harm when people breathed it in. This shows that chlorine can be helpful or deadly, depending on how it is used. That mix of good and bad adds to its strange and powerful image.

Chlorine also appears in stories, films, and symbols. It often stands for purity, cleaning, and change. When something is bleached or cleaned, it can represent a fresh start. This idea has helped chlorine gain a place in art and culture. It is even known as The "Jekyll and Hyde" Element in science literature where it is used as a symbol for dual nature, being extremely poisonous in its gaseous form (used as a chemical weapon in WWI) while being essential for sanitizing water and forming table salt.

In the end, chlorine seems magical because it has strong powers, many uses, and a deep history. It can clean water, help make products, and even protect lives. Yet it can also be dangerous. That balance between beauty, usefulness, and danger is what makes chlorine so fascinating.

Meet Kyrstix, The Dragon with a Magical Tail Tipped with Chlorine

The morning sun peeked over the tall mountains, painting the sky with shades of pink and orange. Krystix woke up slowly. He took a deep breath. The air smelled like fresh rain and sweet pine needles. He stretched his long, powerful body. His bright green scales caught the morning light. They shined like wet leaves in the summer. Krystix was a dragon, but he was not a scary monster. He was a guardian. He loved to learn new things, solve tricky puzzles, and help anyone who was in trouble. His big yellow eyes were full of kindness and smarts.

Krystix lived in a beautiful, quiet forest. The birds sang to him, and the deer were not afraid to walk right past his tail. But on this day, Krystix felt worried. He was thinking about a dark rumor he had heard from the wind. The wind whispered that a terrible sickness was spreading in a valley far away. The rivers there were turning black. The dirt was turning sour. The animals were running away. Krystix knew he had to help. He had a very special kind of magic. He did not breathe fire that burned things down. Instead, he could make a special, glowing green mist called chlorine. This magic could clean up poison and heal the broken land.

Just as he was thinking about this, he heard a crunching sound. Someone was walking on the dry leaves. Krystix looked up and saw a human walking out from the shadows of the tall trees. It was a hunter. Her name was Kira. She wore a thick brown cloak, and she carried a wooden bow on her back. Her boots were covered in gray mud, and she looked very tired. But her eyes were brave.

"I have been looking for you, dragon," Kira said. Her voice was shaking just a little bit. "I need your help. The valley to the north is sick. The land is broken, and the people are in great danger."

Krystix gave her a warm, gentle smile. He lowered his big, horned head so he could look right into her eyes. "Stories can have happy endings, traveler," he said softly. "But sometimes we have to work hard to write those endings. Let us go fix this one together."

Krystix lowered his big green shoulder. "Climb up," he told her.

Kira looked surprised, but she did not wait. She grabbed his thick green scales and pulled herself onto his wide back. She found a safe spot to sit right between two large, smooth spikes on his neck.

"Hold on tight," Krystix warned. With a mighty flap of his giant wings, Krystix jumped into the sky.

Kira gasped. She had never been flying before! Krystix's wings whooshed against the air. They went higher and higher. Kira held on tight as she rode with him on the wind. The cold air rushed past her face, making her cheeks turn red. She looked down and saw the world below. The trees looked like tiny green bushes. The rivers looked like little silver ribbons. It was beautiful.

But as they flew further north, the view changed. The bright green forest faded away. The sky turned gray and cloudy. The air no longer smelled like pine trees; it smelled like rotten eggs and rust. Kira pointed down. "Look! Down there!" she shouted over the loud wind.

Krystix looked down. The valley looked terrible. The river was completely dry. The trees had no leaves, and their branches looked like skinny, gray bones. Worst of all, the ground was covered in a thick, glowing purple slime. It was poison.

Suddenly, the ground started to shake and rumble. It felt like a giant earthquake.

"What is happening?" Kira cried out.

Out from an old, dark mine carved into the side of a mountain, a giant wave of toxic purple mud burst forward. The mine had collapsed! The mud was thick and fast, and it was rushing right toward a group of scared miners. The miners had been trying to dig for shiny rocks, but they had dug too deep. Now, they were running for their lives. They dropped their pickaxes and shovels, screaming as the purple wave chased them down the rocky hill.

"They are not going to make it!" Kira yelled. "We have to do something!"

"Hold on!" Krystix roared. He folded his wings back and dove down fast, like a green arrow shooting from the sky. The wind screamed in Kira's ears. Krystix pulled his wings open at the very last second. He landed with a heavy thud right between the running miners and the deadly wave of mud. Dirt flew into the air.

Krystix spun around to face the danger. He did not open his mouth to breathe fire. Instead, he swung his long, spiked tail around. He pointed the very tip of his tail right at the rushing wave of purple slime.

He closed his eyes and focused. He released his magic through the tip of his tail!

A bright, glowing green mist shot out like a powerful laser beam. The chlorine magic zoomed through the air and crashed right into the toxic mud with a hiss, a bang and a crackle.

The sound was loud, like water hitting a hot pan. The magic fought the poison. The bright green mist wrapped around the thick, ugly mud. It started to break the poison apart. The purple slime bubbled and boiled. Kira watched in amazement as the green magic ate away the bad chemicals. In just a few seconds, the scary, glowing wave of slime turned into a splash of harmless, clear water. It washed gently over the rocks and splashed safely around the miners' dusty boots.

The danger was gone. The miners fell to their knees in the wet dirt. They were breathing hard, covered in sweat, but they were safe. Kira slid off Krystix's back. She held her bow ready, just in case something more sinister than mud came out of the mine, but the mountain was quiet.

Krystix walked slowly over to the miners. He was huge, but he did not try to scare them. He did not yell at them for making a mess. He spoke to them like a wise, patient teacher.

"You dug too deep," Krystix explained, his deep voice echoing in the quiet valley. "You woke up the dark poison that was sleeping in the earth. The land is not just a place to take things from. It is a living thing. You must be careful. If you want to mine here, you must respect the land and keep it clean."

The leader of the miners, a man with a dirty face and a torn shirt, looked up at the great dragon. He nodded quickly. "We are so sorry," the man said. "We only wanted to find crystals to sell so we can feed our families. We did not know the poison was there. We promise to clean up our camp. We will mine safely from now on. Thank you for saving our lives."

Krystix nodded his big head. "I will help you clean the rest of the valley," he promised.

They spent the rest of the day working together. Krystix walked around the valley, using small bursts of magic from his tail to clean the leftover puddles of purple slime. Kira helped the miners pick up their tools and fix their broken tents. By the time the sun started to go down, the valley already looked a little bit brighter.

That night, they all built a large campfire together. Krystix used a tiny, safe spark from his tail to light the dry wood. The fire crackled and popped, sending warm orange light into the dark sky. Krystix and Kira sat together under the stars, a little bit away from the miners.

The air no longer smelled like poison and rotten eggs. It smelled like fresh rain, clean dirt, and the warm smoke of the campfire.

Kira smiled at the big green dragon. She reached out and patted his warm scales. "You did a great job today, Krystix," she said softly. "You saved a lot of people."

"We did a great job," Krystix corrected her. "A guardian always works better with a brave friend."

Krystix rested his heavy head on his front paws. He closed his yellow-green eyes, feeling very tired but very happy. He knew there would always be danger in the world. People would make mistakes. Things would get broken. But as long as he had his magic, his courage, and good friends by his side, he knew he could protect the world and help it heal, one day at a time.

Enjoy This Coloring Page Featuring

Krystix The Dragon with the Chlorine Tipped Tail

Sample Page From Magical Elements of the Periodic Table

Presented By The Elemental Dragons Book

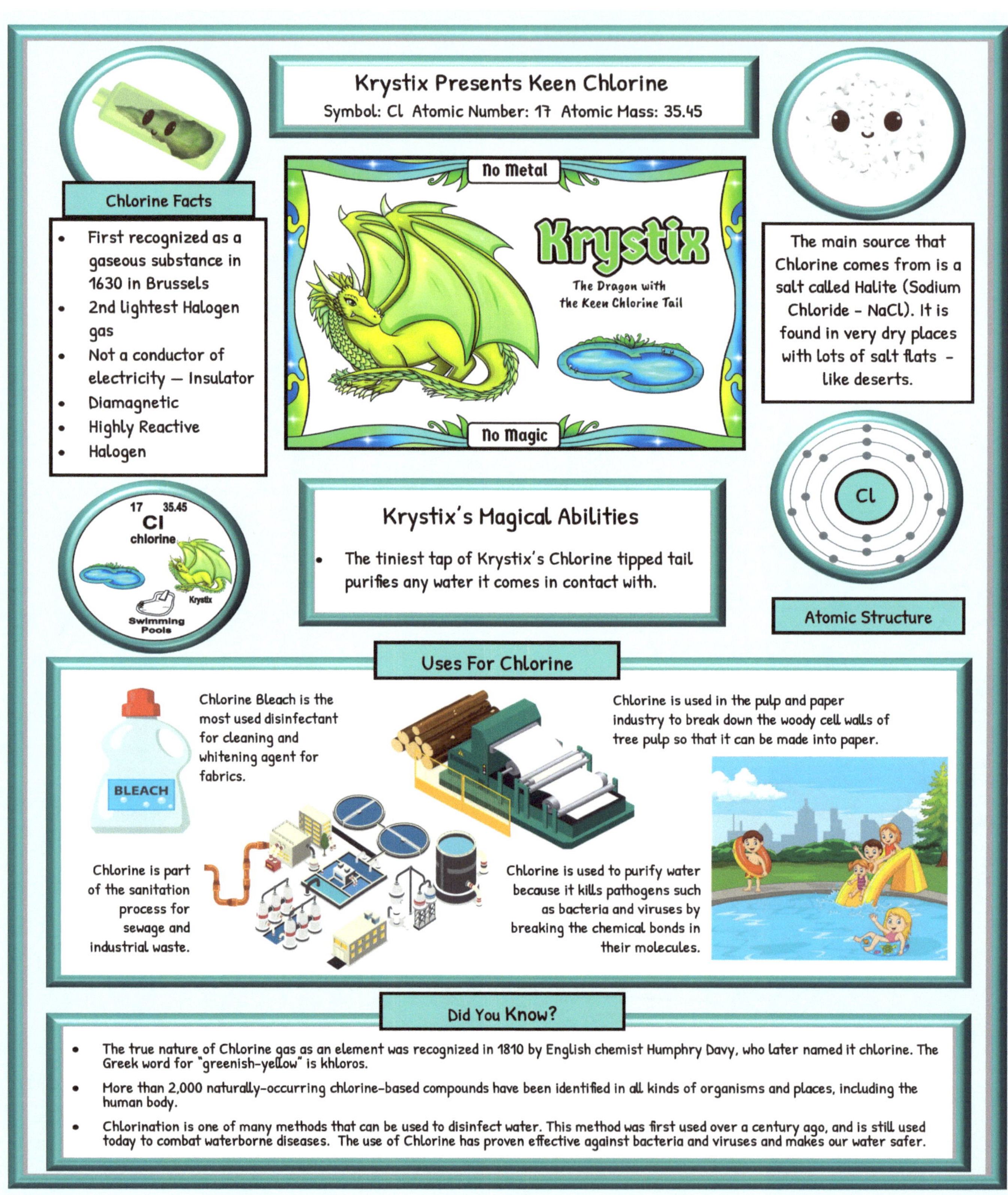

MEET THE CLAN

THE DRAGONS WITH THE MAGICAL ELEMENTAL TIPPED TAILS

Create Your Own Magical Dragon Elemental

Krystix

The Dragon With The Keen Chlorine Tipped Tail

Symbol: Cl Atomic Number: 17 Atomic Mass: 35.45

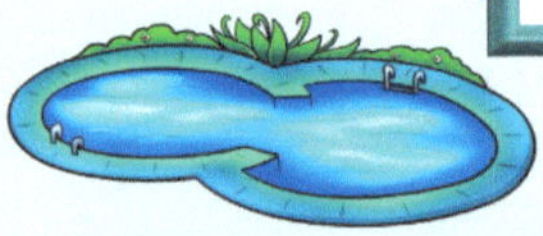

Magical Clan Crest Symbol

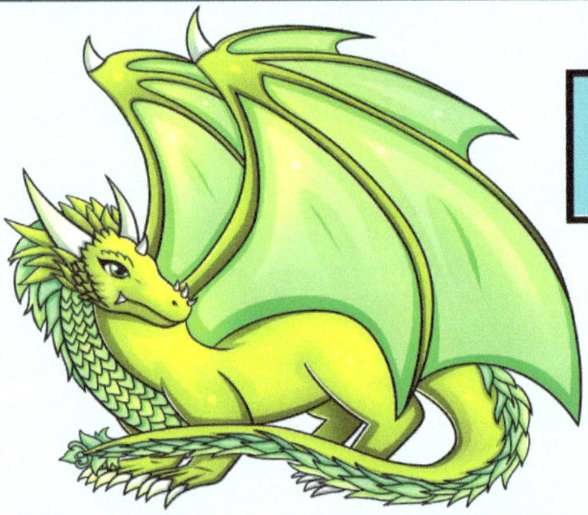

17 35.45
Cl
chlorine

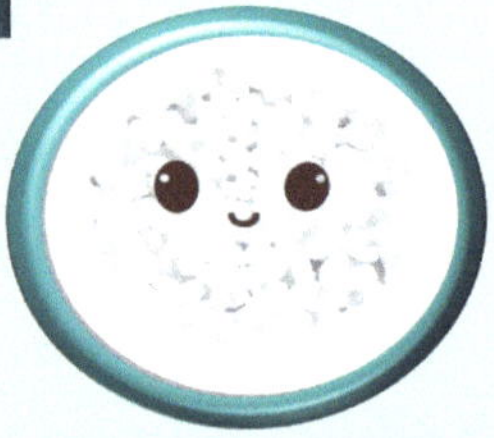

Chlorine is extracted from Halite

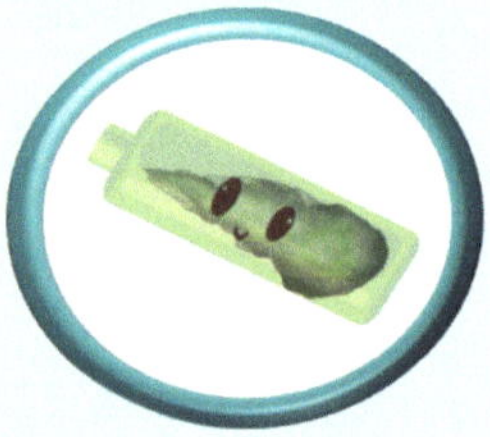

Halogen

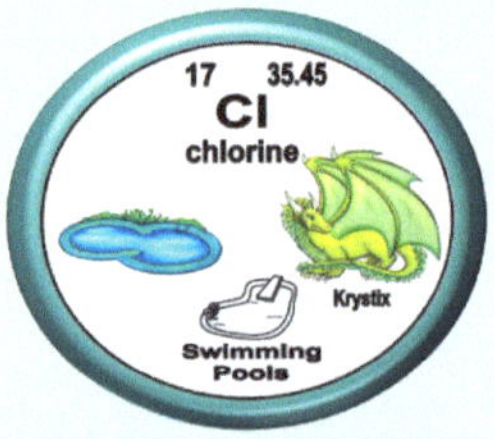

Chlorine Periodic Symbol

Krystix's Magical Abilities

- The tiniest tap of Krystix's Chlorine tipped tail purifies any water it comes in contact with.

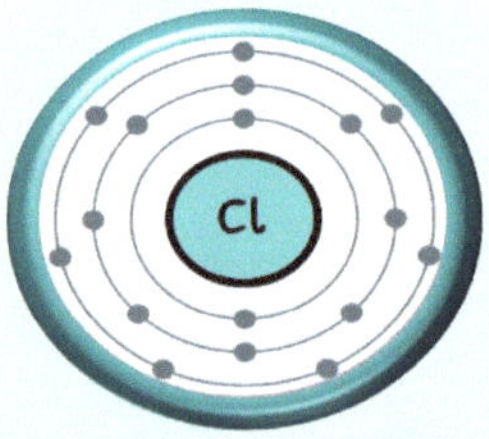

Atomic Structure

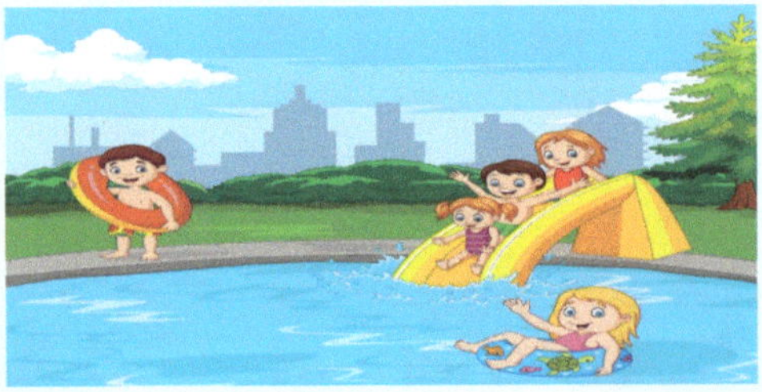

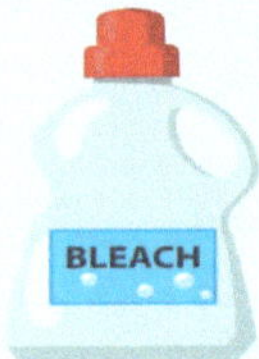

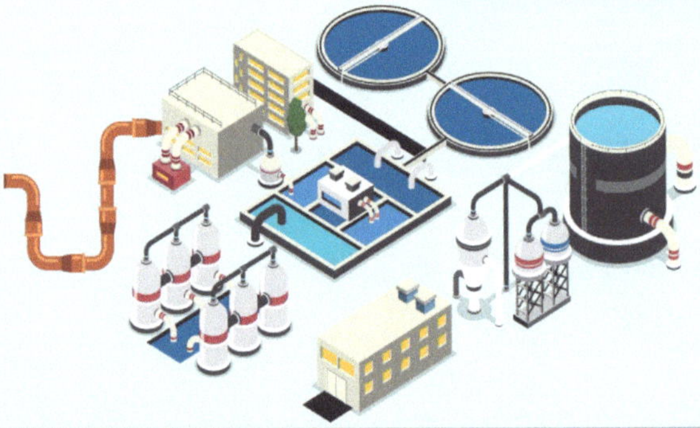

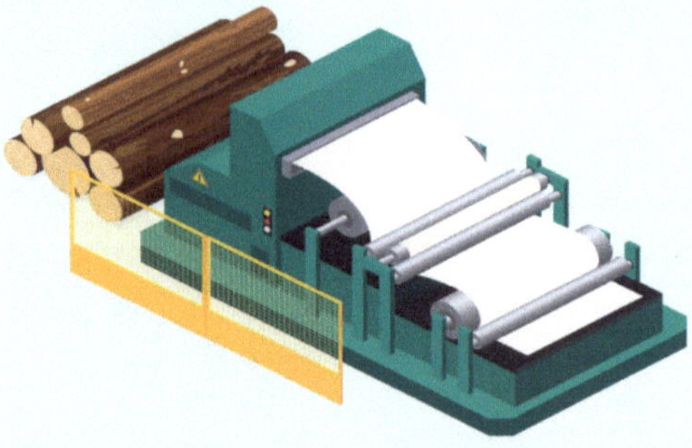

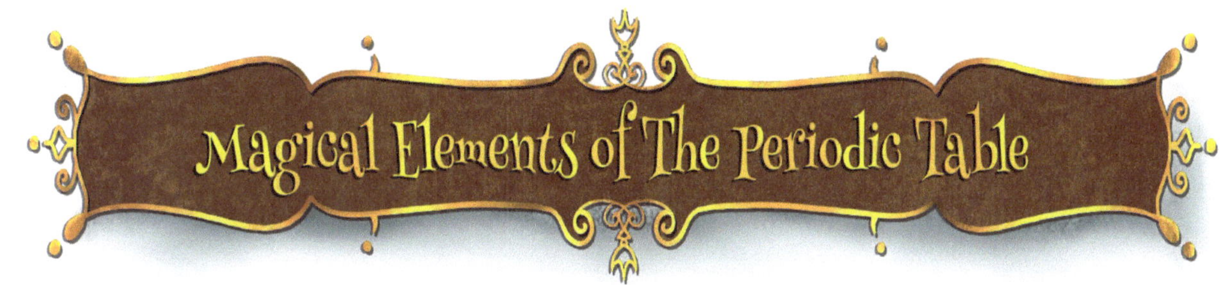

Students may either use a program like power point to cut and paste clip art into a Magical Dragon Elemental Blank or, if they wish, they may draw everything themselves.

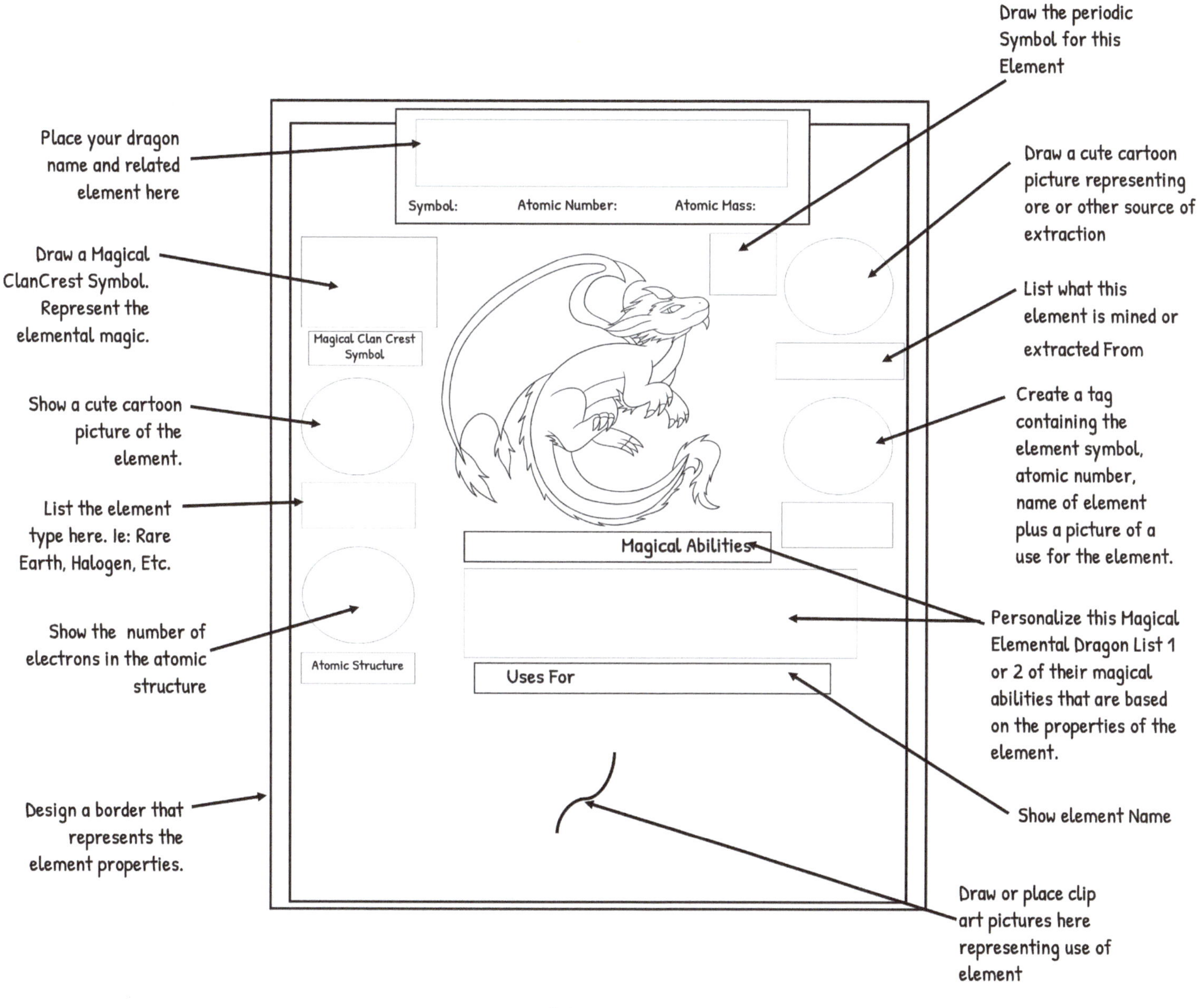

Place your dragon name and related element here

Draw a Magical ClanCrest Symbol. Represent the elemental magic.

Show a cute cartoon picture of the element.

List the element type here. Ie: Rare Earth, Halogen, Etc.

Show the number of electrons in the atomic structure

Design a border that represents the element properties.

Draw the periodic Symbol for this Element

Draw a cute cartoon picture representing ore or other source of extraction

List what this element is mined or extracted From

Create a tag containing the element symbol, atomic number, name of element plus a picture of a use for the element.

Personalize this Magical Elemental Dragon List 1 or 2 of their magical abilities that are based on the properties of the element.

Show element Name

Draw or place clip art pictures here representing use of element

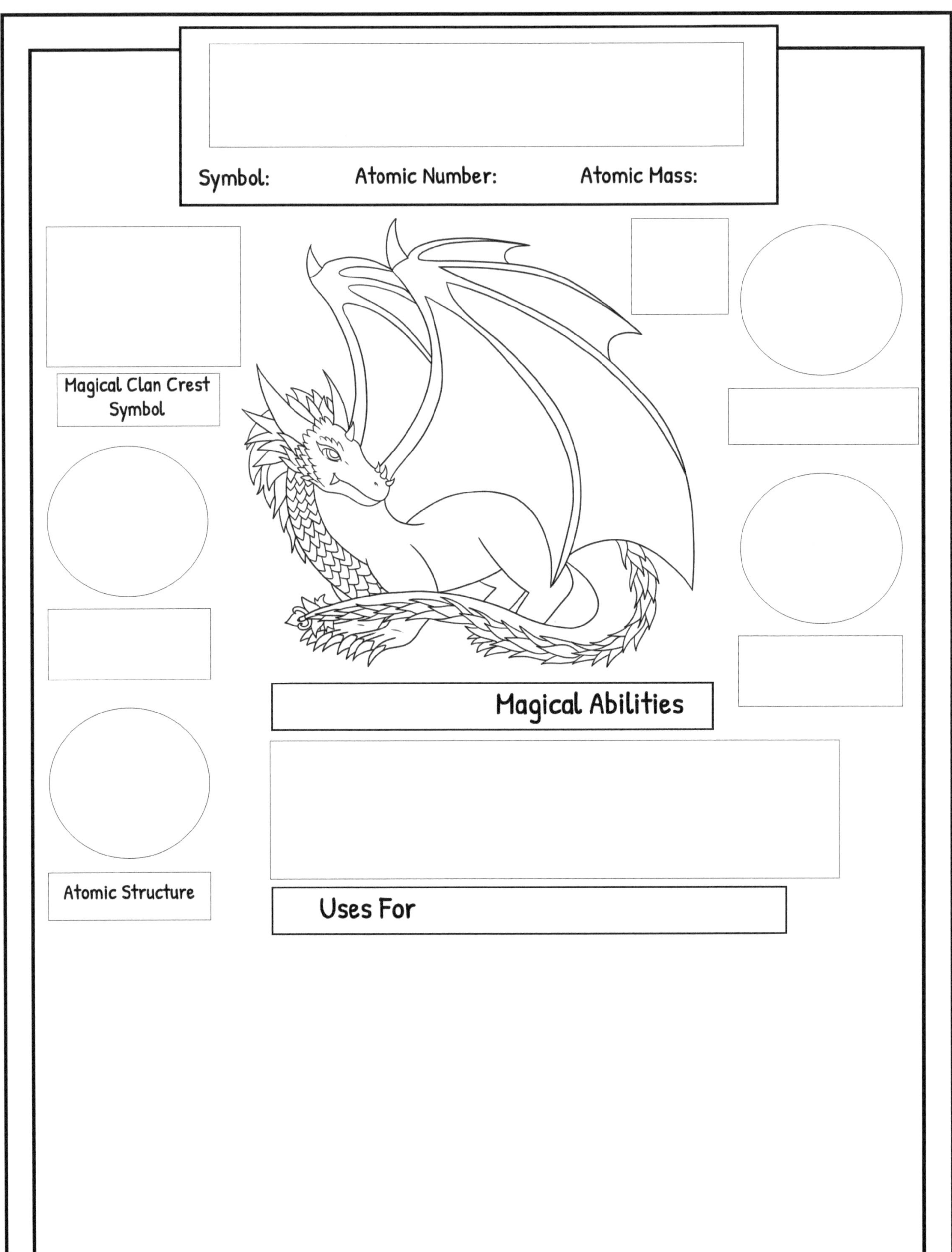

Symbol:
Atomic Number:
Atomic Mass:
Magical Clan Crest Symbol
Magical Abilities
Uses For
Atomic Structure

Magical Dragon Elemental Research Sheet

Before starting your Magical Dragon Elemental graphics page, do some research on your chosen element.

Name of Magical Dragon:	
Dragon's Magic Power Based on the Element's Properties:	
Magical Clan Crest Symbol:	
Element Name:	
Element Symbol:	
Atomic Number:	
Atomic Mass:	
What year and where was this Element discovered?	
Who discovered this Element?	
Element Group:	
Element Period:	
Element Family Name:	
State of Element At Room Temperature:	
What is Element Mined or Extracted From?	
Is Element Magnetic?	
Does Element Conduct Electricity?	
Where is the Element commonly found in Nature?	
What is 1 alloy of the Element? How used?	
What is 1 compound of the Element? How used?	
Name the most common use for this Element:	
Name a little known use for this Element:	
Name one more use for this Element:	
Interesting and Fun Facts:	

Magical Unicorn Elemental Research Sheet

Before starting your Magical Unicorn Elemental graphics page, do some research on your chosen element.

Name of Magical Unicorn:	Ghel The Gold Horn Unicorn
Unicorn's Magic Power Based on the Element's Properties:	Ghel can see past, present and future. She is empathic and can sympathize with the feelings of other. They say she has a heart of gold.
Magical Herd Crest Symbol:	An open heart with a Celtic Trinity Knot.
Element Name:	Gold
Element Symbol:	Au— Comes from Aurum which is the Latin word for Gold.
Atomic Number:	79
Atomic Mass:	196.97
What year and where was this Element discovered?	Around 4,600 BCE in Bulgaria
Who discovered this Element?	Unknown
Element Group:	11 on Periodic TAble
Element Period:	6 on Periodic TAble
Element Family Name:	Gold is a Noble Transition Metal
State of Element At Room Temperature:	Solid
What is Element Mined or Extracted From?	Quartz Veins. It is also found in gravel in streams.
Is Element Magnetic?	It is Diamagnetic. It's only weakly magnetized when placed in a magnetic field.
Does Element Conduct Electricity?	Gold is a great electrical conductor used in printed circuitry of computers.
Where is the Element commonly found in Nature?	One of the largest deposits is found in the United States in Arkansas.
What is 1 alloy of the Element? How used?	White gold is an alloy of gold, palladium, nickel and zinc.
What is 1 compound of the Element? How used?	Gold Phosphide is a semiconductor used in high power, high frequency applications and in laser diodes.
Name the most common use for this Element:	Jewelry
Name a little known use for this Element:	Acupuncture needles
Name one more use for this Element:	Gold is used in airbags in cars.
Interesting and Fun Facts:	Gold was used in ancient Egypt to fill decayed teeth. Gold thread is incorporated in astronaut spacesuits to protect them from the heat of the sun.

Write a paragraph below to describe your magical dragon elemental. Based on the information obtained from research of your chosen element, how did you determine your dragon's name? What are your dragon's magic powers? What are their likes/dislikes, strengths/weaknesses, personality traits? What color is your dragon and why did you pick that color? What is your dragon's Clan Crest Symbol?

Magical Unicorn Elemental Sample Description

Ghel The Gold-Horned Unicorn

This magical unicorn has a golden horn and hooves that glow like the sun. Her hide is honey-gold and her flowing mane and tail are golden-blonde.

Ghel is a member of the Metal Horn Unicorn Tribe from Unimaise. Gold is linked to the heart chakra because it holds a warm energy that brings soothing vibrations to the body to aid in the healing process. Ghel's Magical Herd Crest symbol is an open heart with a Celtic Trinity knot which indicates that she can see past, present and future. She is empathic and can sympathize with the feelings of others.

Being empathic does not make Ghel a weakling. She is brave with strong opinions and is a true champion to those she loves. She is known as the unicorn with the "heart of gold". When she places her horn on the heart of another, she senses their future.

The name Ghel is an Indo-European word which means yellow. The word "gold" most likely has its origins in the word "Ghel".

Gold is known to possess spiritual powers that bring happiness, peace, stability and luck to those who wear it. Scientists say that all the gold in the world comes from the collision of neutron stars.

Though most other magical unicorn elementals get along well with the gold horn unicorn herd; Ghel, and others like her, must be very careful around the Quick Silver Herd - as gold dissolves in mercury.

Do Your Middle Graders Want To Know More From The Magical Elementals About The Periodic Table?

Get the accompanying books in print at all online book stores. Get the books and accompanying activities at MagicalPTElements.

Available Now

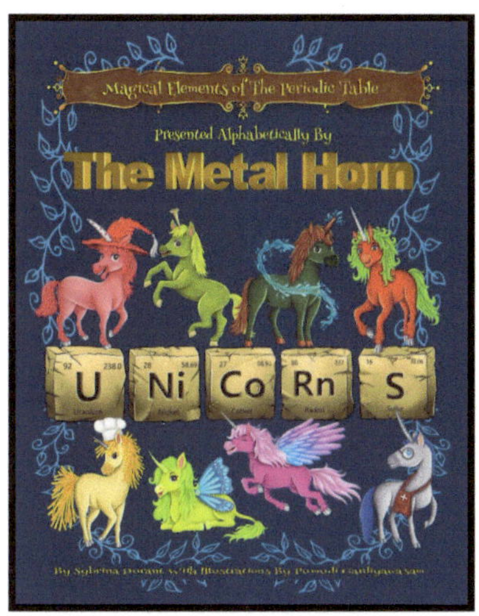

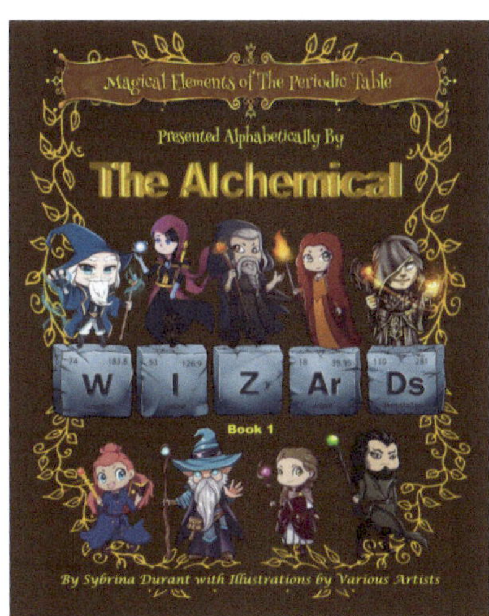

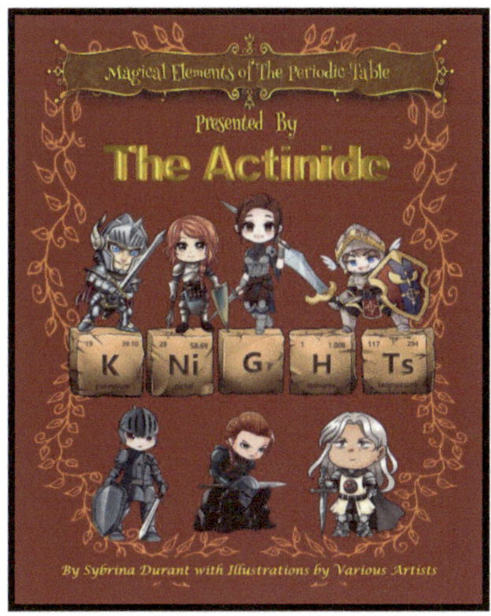

Get These Trading Cards Sets Featuring The Magical Elementals Representing The Periodic Table Elements

at https://bit.ly/4OoEUBr

Unicorns, Dragons, Wizards, Knights, or Goblins?

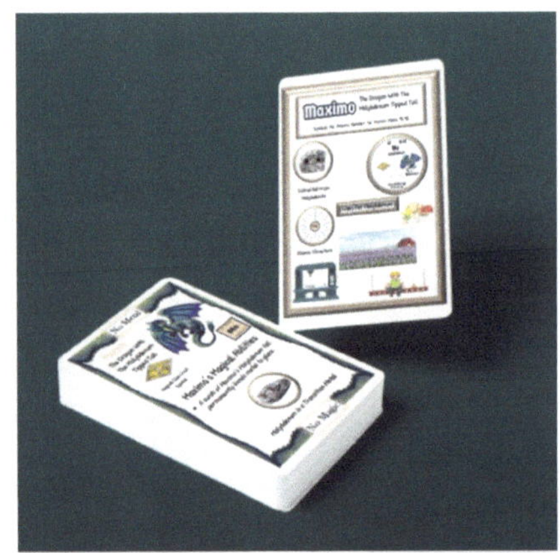

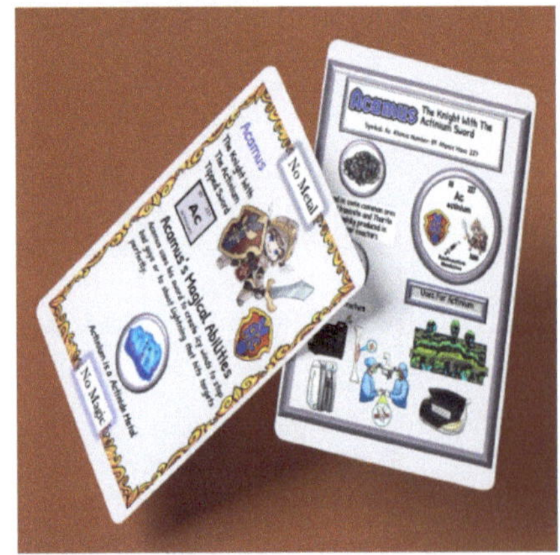

Collect Them All

These magical elementals are ready to help make learning the periodic table more fun. Get all books sets related activities today.

Learn More About all of the Periodic Table Elementals.
Get all of the "No Metal No Magic" Books Featuring
Individual Elements at MagicalPTElements.com

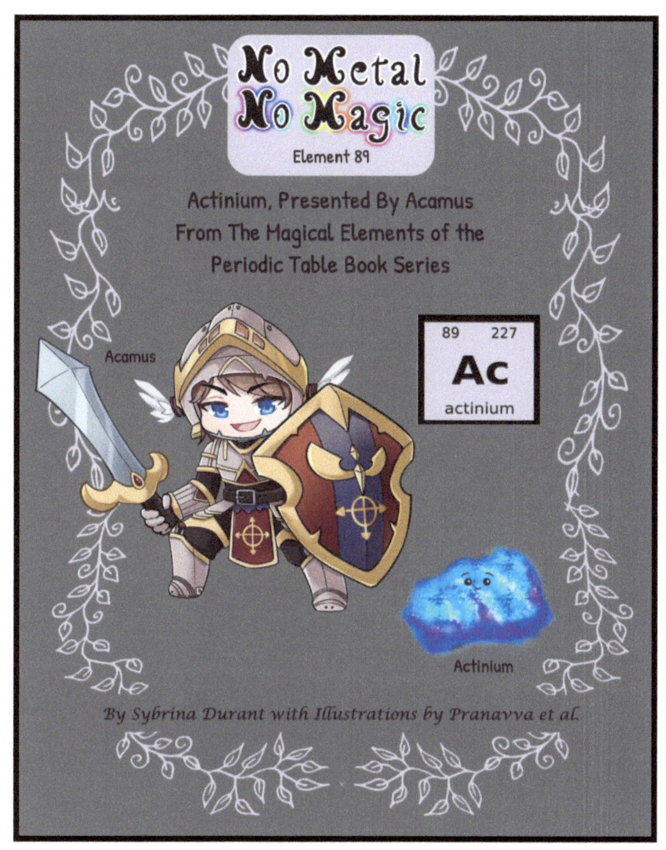

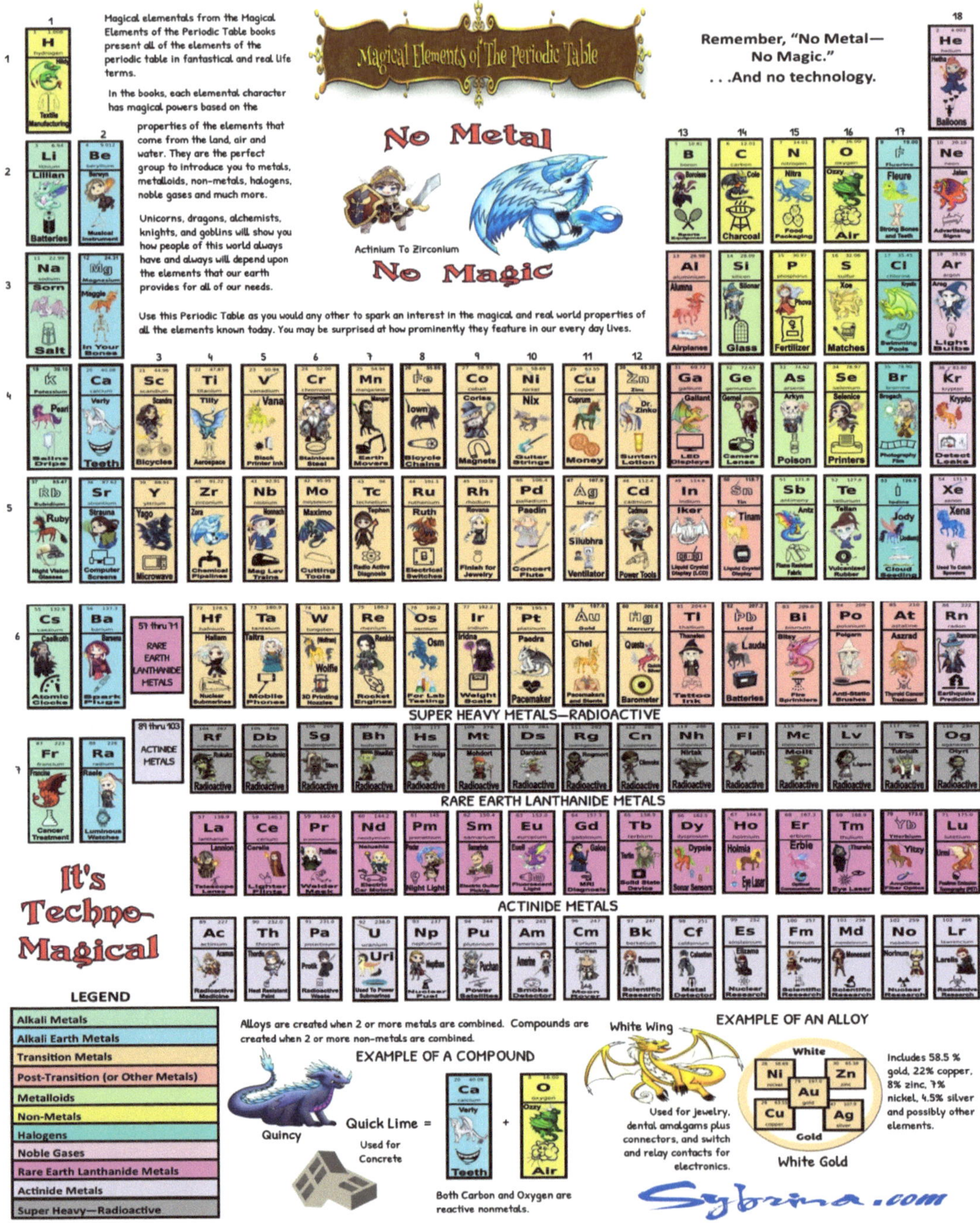

Would you like a 24" x 36" poster of the Magical Elemental-Themed Periodic Table from The Magical Elements of the Periodic Table books? The best place to get a high quality poster size print is at https://bit.ly/49QMxBT It will look great on a classroom or kid's room wall.

Get These Fun Elemental Periodic Table Activities at

MagicalPTElements.

Unicorn Periodic Table Bingo—Comes with 32 unique Bingo cards. Magical Elementals Bingo comes with 36.

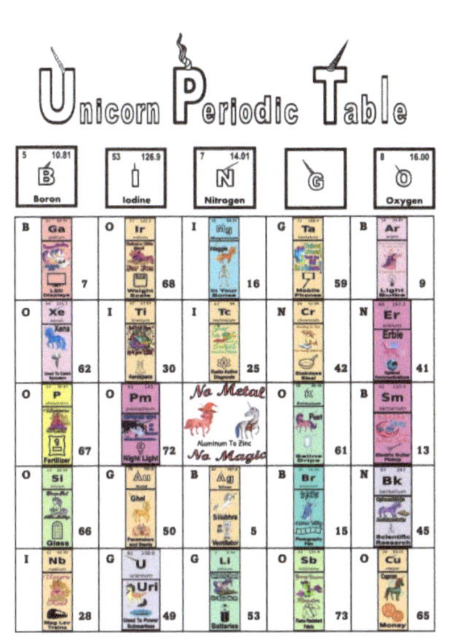

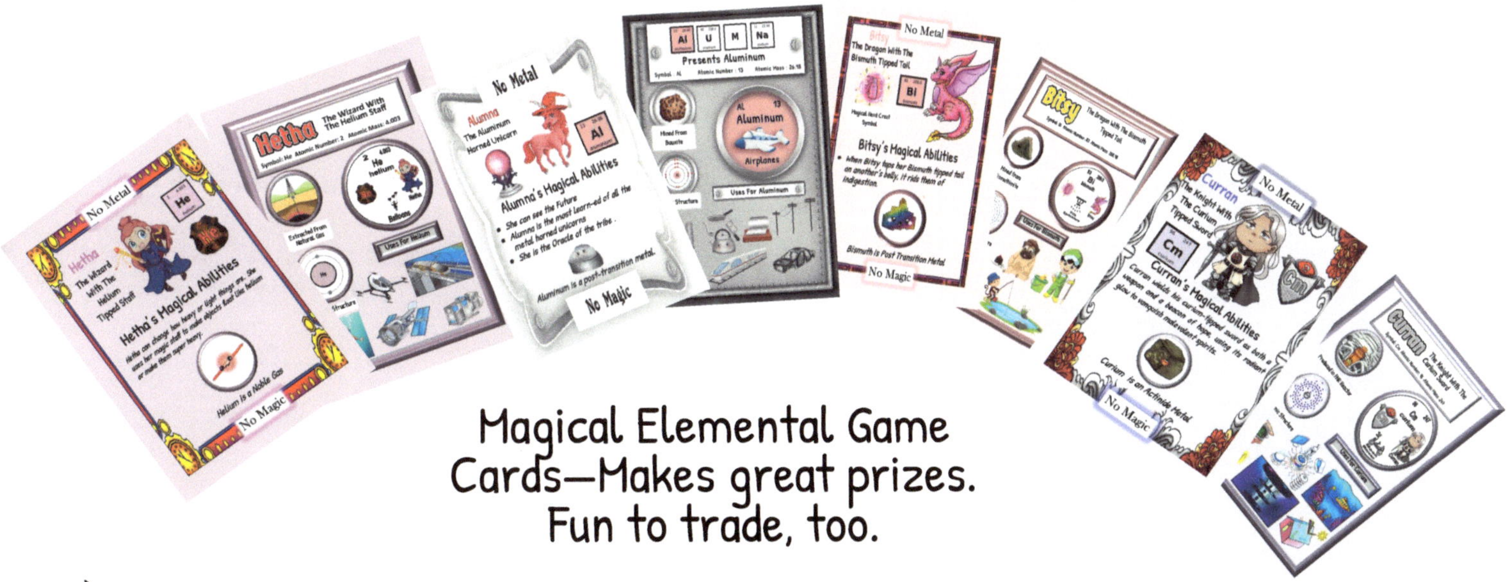

Magical Elemental Game Cards—Makes great prizes. Fun to trade, too.

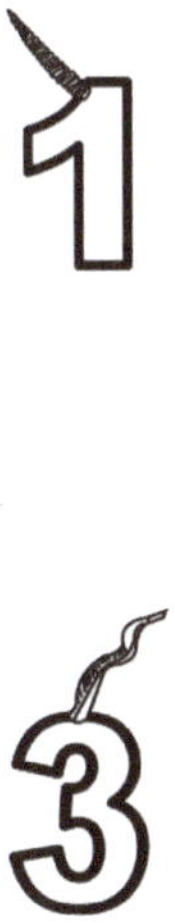

plus

1 2 3

U nicorn H orn

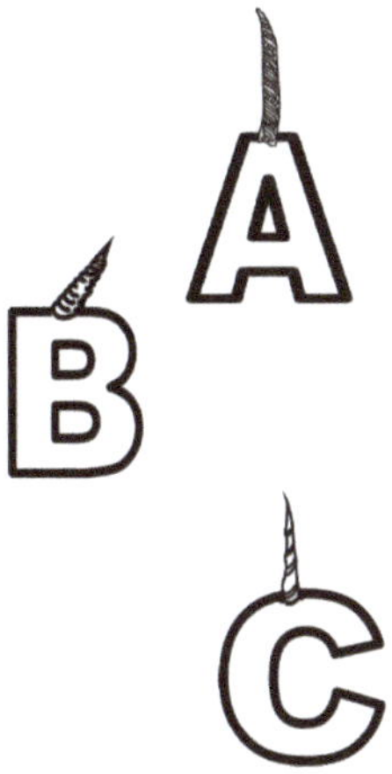

A lphabet

A B C

Also browse activities at
https://www.magicalPTelements.com
for all kinds of printable downloads to make learning fun.

Printable Magical Elemental Activity Downloads

Fun Way For Students To Learn The Elements Of The Periodic Table

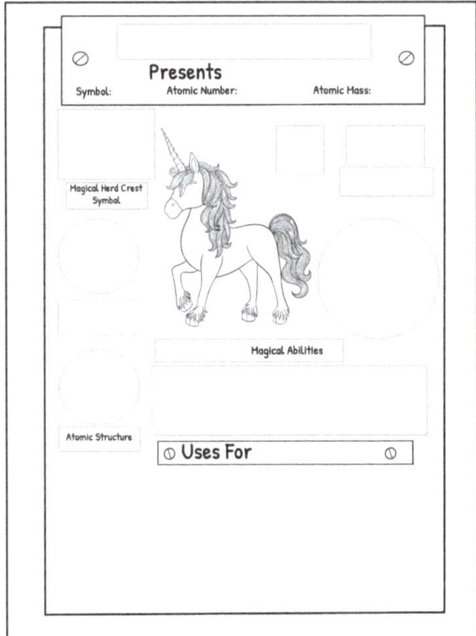

Blank Unicorn Element Card

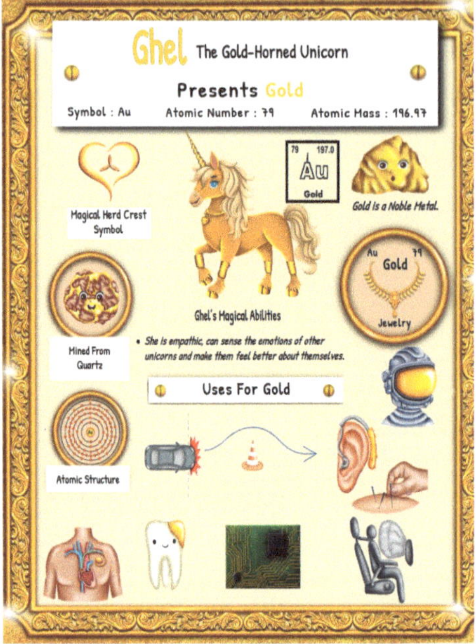

Sample Unicorn Element Card

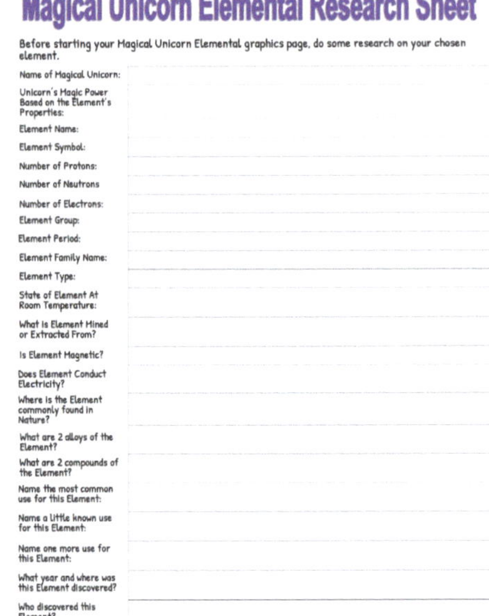

Blank Research Sheet

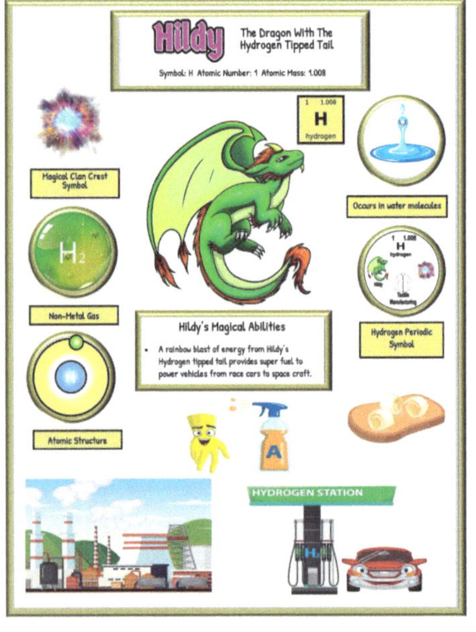

Sample Dragon Element Card

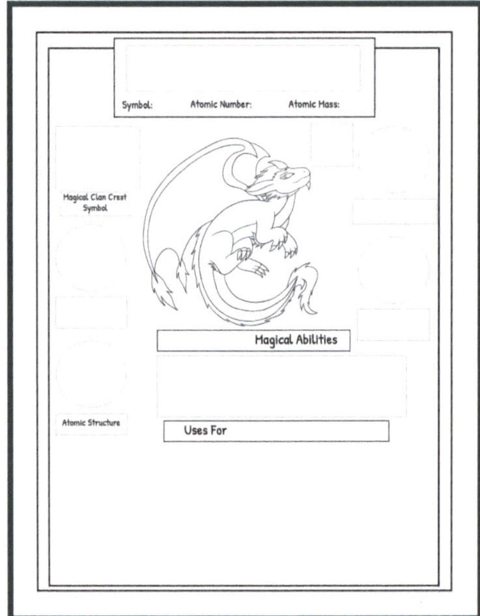

Blank Dragon Element Card

Blank Research Sheet

Using the sample Magical Elemental cards provided, have students select an element from the Periodic Table and a Magical Elemental Card Blank to create their own Magical Elemental Card. The blank and sample cards do not have to match.

You will receive a pdf containing either 26 unicorn or 26 dragon sample cards and blanks to be printed on 8 1/2 x 11 sized paper or card stock. The pdf also contains a Magical Elemental Research Sheet for the students to work on before creating their unique Periodic Table Elemental. They will also write a short paragraph describing their Unicorn or Dragon Elemental from that research.

Get These Fun Elemental Periodic Table Activity Sheets at MagicalPTElements.com

Don't Forget To Get A Tee Shirt

Featuring Your Favorite of the

118 Elements from the Periodic Table

Available in adult and kid sizes in many colors.

https://amzn.to/47NVZWN

This is the Chlorine-Krystix Tee Shirt Graphic

Get it at https://www.amazon.com/dp/B0D3Y5KW8V

Dear Reader

I hope the "No Metal No Magic Element 17 — Krystix, from
The Magical Elements of the Periodic Table Series Presents
Helium" with illustrations by Pranavva et al, has helped you
learn some fun and interesting things about the magic of the
element, Chlorine.

This is one of what will eventually be 118
books featuring periodic table elements
presented by unicorns, dragons, wizards,
knights and goblins. Keep checking
regularly. Every one of the elements are
amazing and very necessary to our

Techno-magical.

A lot of research went into every page of this book as well as the Magical Elements of the
Periodic Table Books. There are just too many references to publish in this book but you can
read and research them all at MagicalPTElements.com/MAUPT or /MDAPT or /MW1PT or /
MW2PT or /MAKPT or /MRGPT. There, you can also access book related activity sheets and
games to help make the learning process more fun.

Get ready made trading cards, lapel pins, tee shirts and more based on this book from Sybrina
Publishing's No Metal No Magic Collection at Zazzle - **http://bit.ly/3km64Wg**

Would you like a 24" x 36" poster of the Elemental-Themed Periodic Table in this book? The best
place to get it is at **https://bit.ly/49QMxBT** They have the sharpest images of any other
poster printer around.

The Magical Elements of the Periodic Table books came into existence because of

my Blue Unicorn—Journey To Osm books. If it weren't for their magical powers,

based on the properties of the metals of their horns and hooves, I would have

never come up with the idea to relate magical creatures to the periodic

table. There's a metal horn unicorn story for every age group and

they are all available at MagicalPTElements.com

If you enjoyed this book
please leave a nice review
at your favorite online book site.

No Metal No Magic

Song Lyrics

No metal, no Magic

No metal, no Magic

I can think of nothing more tragic

Than to have no metal or no magic

Metal makes everything magical.

Just ask a unicorn. . .

Preferably, one with a metal horn.

They'd say No metal, No magic.

Metal makes everything techno magical.

No metal, No magic

for two-leggers or unicorns.

No metal, No magic

Metal makes everything techno magical.

No metal, No magic

It's techno magical.

No metal, No magic

It might be very hard to believe but with

No metal, No magic

There'd be no technology.

No metal, No magic

Blue Unicorn

Ebooks, Audio and Print Books Available at all online book stores.

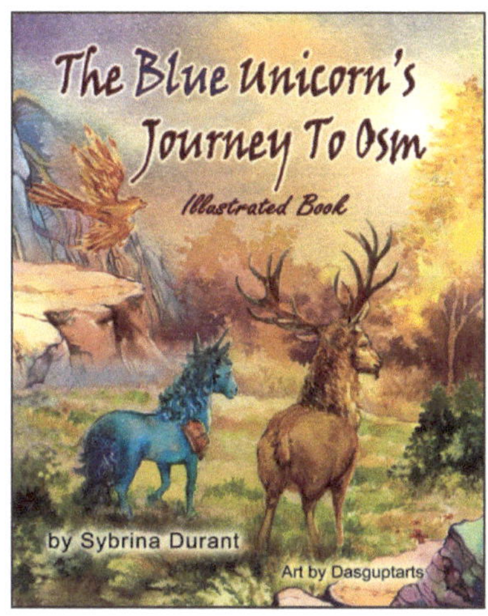

Illustrated Book

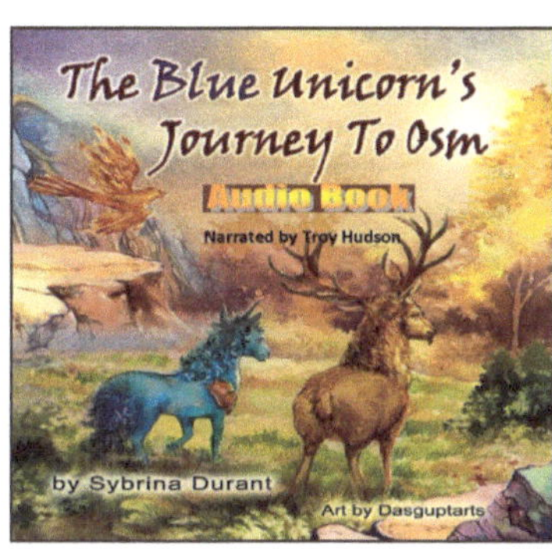

Audio Book

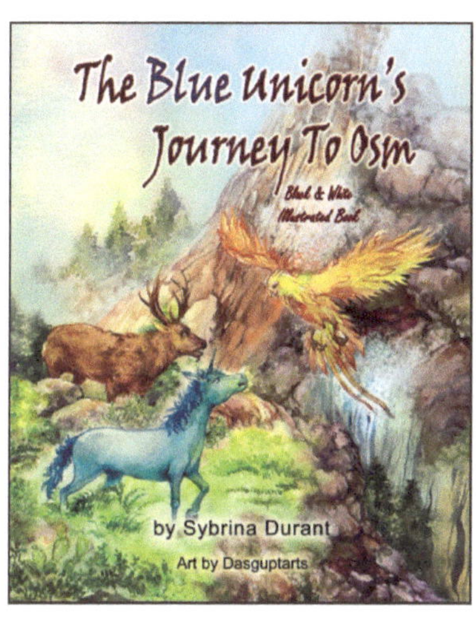

'Read & Color' Book For Teens

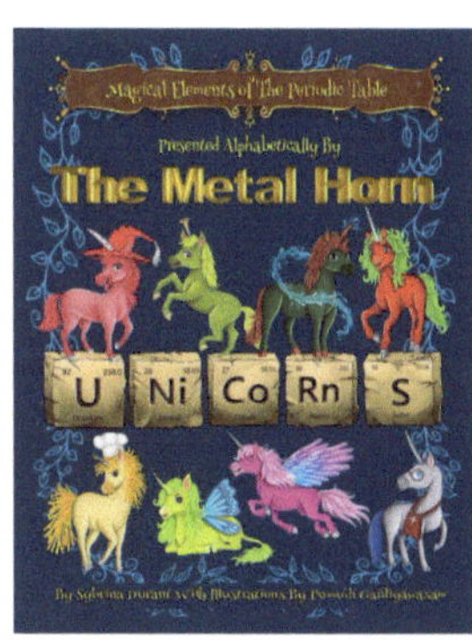

Unicorn Periodic Table Book

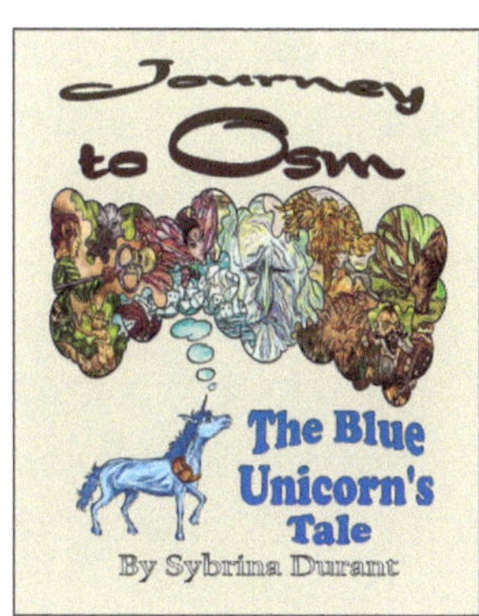

Fantasy Novel

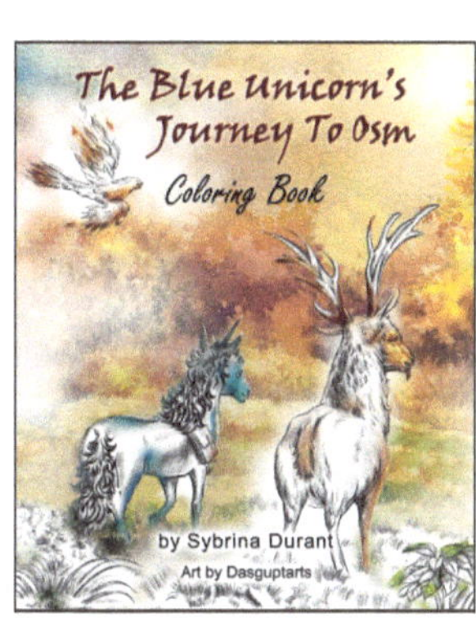

Coloring Book & Character Introduction

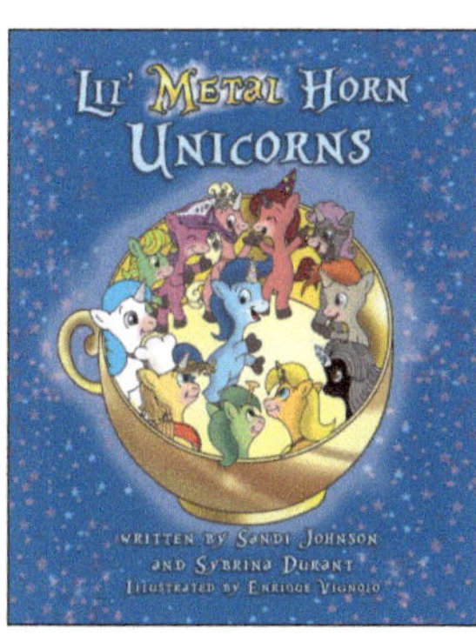

Picture Book For Kids

Get these and more at
MagicalPTElements.com
and Sybrina.com

Back then, most places throughout MarBryn had wizards and sorcerers of some ability, or another. Most were trained in the ways of magical arts by the unicorns as part of their outreach program. Some two-leggers developed practical magical skills like making delicious feasts of tasty food appear out of thin air or purifying murky water around the land.

Others went through more extensive training to learn battle magic—like shooting powerful streams of energy from their swords.

Some were taught the art of holding the glow of the sun in magical globes, bringing light into the dark of night. These magical lights warded off the evil beings that were new and frightening products of dark magic.

With the rise of the sorcerer Magh, magical defense arts had become more important. The highest level of magical training involved sensing when others were in danger and learning to see into the future. Very few two-leggers ever reached that level because magic wasn't inherent in them the way it was for the unicorns. The metal of their horns and hooves were part of them as well as the very makeup of their blood, but two-leggers relied on learned magic via potions, charms, and incantations that required help from ingredients and forces more mystical in nature than any two-legger was ever born to be. Of course, controlling magic and projecting your intentions went far beyond merely following a recipe of sorts. It took being in touch with nature and the various elements to get the response a wizard desired. Much trial and error went into it, as well as faith and trust and the motives of the spell caster.

Magic was and is a practice that is never quite perfected even for the unicorns who must continue to hone and learn how to harness their powers. On rare occasion a wizard and unicorn had formed enough trust and a steadfast bond that prompted the unicorn to gift the wizard with a wand or staff embedded with the smallest sliver of metal from one of their hooves. This was rare but had happened and of course so had the desire for more power. Magh wasn't the first sorcerer with lust for more and throughout history there had been a handful of heinous acts against unicorns from those seeking their magic. Prior to Magh those wizards had failed to circumvent the protections nature had infused unicorn magic with so even after harvesting metal from their horns or hooves these sorcerers had gone mad trying to bend the will of nature and actually use their ill-gotten gains.

Magh, too, had gone mad or perhaps he'd already been so, but somehow he'd managed to harness the metal he harvested from his victims and continued to grow stronger rather than completely lose his mind like the others. How exactly remained a mystery to the unicorns and everyone else.

The wizards who'd honed their craft with the help and blessing of the unicorns tried to solve the riddle and even the best of them failed to uncover that secret. Some of MarBryn's natives took to magic naturally, while others struggled with the concept but no matter how adept they were. All of them fought valiantly against Magh's magic because with even a shred of knowledge they understood how the power shift would ultimately play out. They fought to the end but, in the end, only one sorcerer remained in MarBryn and now, Magh was in total control. But still, he was not satisfied. He wanted to control every living creature in the land.